CRYSTALLOGRAPHIC AND STRUCTURAL ANALYSIS OF IRON CARBONYLS

Crystallographic and Structural Analysis of Iron Carbonyls

Milan Melnik, Iveta Ondrejkovičová,
Vlasta Vančová, Mária Kabešová
and Clive E. Holloway

Nova Science Publishers, Inc.
Huntington, New York

Editorial Production:	Susan Boriotti
Office Manager:	Annette Hellinger
Graphics:	Frank Grucci and Jennifer Lucas
Information Editor:	Tatiana Shohov
Book Production:	Donna Dennis, Patrick Davin, Cathy DeGregory, and Lynette Van Helden
Circulation:	Latoya Clay, Anna Cruz, and Lisa DiGangi

Library of Congress Cataloging-in-Publication Data
Available upon request

ISBN 1-56072-827-2

227 Main Street, Suite 100
Huntington, New York 11743
Tele. 631-424-6682Fax 631-424-4666
e-mail: Novascience@earthlink.net
e-mail: Novascil@aol.com
Web Site: http://www.nexusworld.com/nova

Printed in the United States of America

Crystallographic and Structural Analysis of Iron Carbonyls

Milan Melnik, Iveta Ondrejkovičová,
Vlasta Vančová, Mária Kabešová
and Clive E. Holloway

Nova Science Publishers, Inc.
Huntington, New York

Editorial Production:	Susan Boriotti
Office Manager:	Annette Hellinger
Graphics:	Frank Grucci and Jennifer Lucas
Information Editor:	Tatiana Shohov
Book Production:	Donna Dennis, Patrick Davin, Cathy DeGregory, and Lynette Van Helden
Circulation:	Latoya Clay, Anna Cruz, and Lisa DiGangi

Library of Congress Cataloging-in-Publication Data
Available upon request

ISBN 1-56072-827-2

227 Main Street, Suite 100
Huntington, New York 11743
Tele. 631-424-6682Fax 631-424-4666
e-mail: Novascience@earthlink.net
e-mail: Novascil@aol.com
Web Site: http://www.nexusworld.com/nova

Printed in the United States of America

Contents

Legend To The Figures

0. ABBREVIATIONS

[14]annulene	7,16-dihydro-6,8,15,17-tetramethyldibenzo[b,i][1,4,8,11] tetraazacyclotetradecimate
anpedt	cis-1-acetyl-2-(p-nitrophenyl)ethylene)dithiolate(2-)
$(AsMe_2)_2C_6H_4$	mono-o-phenylenebisdimethylarsine
asp	1-(dimethylarzino)-2-(diphenylphosphino) tetrafluorocyclobutene
atpd	1-aza-2,3,5-triphosphapenta-1,4-diene
$B(4\text{-}Bu^tcat)$	4-tert-butylcatecholborate(2-)
bcgdt	bicyclo[2.2.1]gent-5-en-2,3-ylendithiolate
bmpa	1-tert-butyl-3-methyl-2-phenyl-η-1,2-azaborolinyle
bmptea	bis{2-[(mercaptophenyl)thio]ethyl}amine
bmptee	bis{2-[(2-mercaptophenyl)thio]ethyl}ether
bpy	2,2'-bipyridine
Bu_4^t-dttd	(tetra-t-butyl)-2,3:8,9-dibenzo-1,4,7,10-tetrathiadecane(2-)
$Bu^tPS_2PBu^t$	tio-bis(t-butylphosphide)
$Bu^t_3(SiMe_3)N_2hx$	1,2,6-tri-tert-butyl-5-(trimethylsilyl)-3,4-diazatricyclo $[3.1.0.0^{2,6}]$ hex-3-en
$(CF_3)_2PNPPh_3$	diphosphazene
$(CH_2NMe)_2B_2S_2$	boranedithiolate
$CH_3N(PF_2)_2$	methytlaminobis(difluorophosphine)
$C_2B_9H_{11}$	carborane(2-)
C_2-cap	capped pophyrine
C_3-cap	a capped pophyrin with four $-(CO)O(CH_2)_3O$- linkages connecting the 1,2,4,5-positives of a benzene cap to the ortho positions of tetraphenylporphyrinate (2-)
cct	cis,cis,trans
chdt	1,2-cyclohexanedithiolate
$C_5H_{10}N_2$	$CN(Me)(CH_2)_2NMe$
C_4H_8S	tetrahydrothiophene
$C_4H_4N_2$	1,2-diazine
C_6H_4SNH	o-aminothiophenol
$C_6H_{11}S$	cyclohexylthiolate(1-)
$C_{50}H_{54}N_8$	substituted octaaza[14] annulene
$C_{31}H_{48}O_2P_2$	di(2,6-tertbutyl-4-methyl-phenolato phosphine)methane
$C_{31}H_{48}P_2$	di(2,4,6-tri-isopropylphenylphosphine)methane
$C_{33}H_{68}B_4N_4P_2$	derivate of 1,1,2,2-tetraborylethane
$C_6Ph_4S_2$	1-(diphenylcyclopropenyliden)-2,3-diphenyl-2-propen-1,3-dithiolate
cp'	tetramethylcyclopentadienyl
crypt	$N(CH_2CH_2OCH_2CH_2OCH_2CH_2)_3N$
2,2,2-crypt	2,2,2-cryptate
cy	cyclohexyl
cyclo-1,3-$C_4H_8S_2$	cyclo-1,3-dithiane(cyclo 1,3-dithiocyclohexane)

dab	2,3-diazabornene
ddpp	5,15-[2,2'-dodecanediamido)-diphenyl]:α,α-10,20-bis(o-pivaloylaminophenyl)porphyrin
dent	deuteroporphinate
dhs	N,N'-dehydrose-midinate
diars	o-phenylenebis(dimethylarsine)
dio	1,4-dioxane
dmda	5,7-dimethyl-3H-1,2-diazepine
dmf	dimethylformamide
dmgBBN	borylated dimethylglyoximate(1-)(BBN- 9-borabicyclo [3.3.1] nonane)
$dmgBH_2$	borylated dimethylglyoximate(1-)
dmsp	2,4-dimethyl-4-sulfido-2-penten-3-olate
dpe	bis(diphenylphosphino)ethylene
dppe	1,2-bis(diphenylphosphin)ethane
dppea	N,N-bis(diphenylphosphino)ethylamine
dppee	1,1-bis(diphenylphosphino)ethene
dpph	hexamethylene bis(diphenylphosphine)
dppm	bis(diphenylophosphino)methane
dppp	bis(diphenylphosphino)pentane
dpttd	2,3,11,12-dibenzo-14,7,10,13-pentathiatridecane(3-)
dsf	1,4-diselenefulvene
dttd	2,3,8,9-dibenzo-1,4,7,10-tetrathiadecane(2-)
edtbt	2,2'-(ethylenedithio)dibenzothiolate
enbt	2,2'-ethylene-bis(nitrilomethylidene)bis-benzenethiolate(2-)
eph	diphenyl-N,N'-dimethylephosium(1+)
2-Etim	2-ethylimidazole
$Et\text{-}S_4$	ethyl-1,2-bis(mercaptophenylthio)ethane(2-)
Et_2S_5	diethyl-2,2'-bis(2-mercapto-phenylthio)diethylsulphide(2-)
Et_2S_4	diethyl-1,2-bis(mercaptophenylthio)ethane(2-)
fdt	2,2-tetrahydrofurylen-dithiolate
fmap	2-fluoro-1,3-dimethyl-1,3,2-diazaphospholidine
$HB(pz)_3$	hydridotris(3,5-dimethyl-1-pyrazolyl)borate(-1)
hdt	bicyclo[2.2.1]hept5-ene-2,2-ylenedithiolate
hmpa	hexamethylphosphoric triamide
hpfb	1,2-bis(dicyclohexylphosphino)-3,3,4,4-tetrafluorocyclobut-1-ene
htcgd	tricyclo[2.2.1.$0^{3,5}$]hept-2,6-ylendiothiolate
hz	hydrazine
mcpp	3,3-bis(methoxycarbonyl)-4-phenyl-1-pyrazoline
mdt	methane-dithiolate(2-)
mftdm	2,2'-bis(2-mercaptophenylthio)diethylmethylamine(2-)
1-Meim	1-methylimidazole
$(Me_2NB)_2S_2$	bis(dimethylamino)diborane(4)-B,B'-dithiolate
$MeN(PF_2)_2$	methylaminobis(difluorophosphine)
$(mesB\text{-}PBu^t)_2$	1,3,2,4-diazadiboretane

Me_2S_4	dimethyl-1,2-bis(mercaptophenylthio)ethane(2-)
MePhsic	2-methyl-4-neopentyl-3-phenyl-1-silenacyclobut-2-ene
$(Me_2PHO)_3P$	tris-(3,5-dimethylphenyl)phosphite
$Me_3SiCH(CH_2)S_2$	1-trimethylsilyl-S,S-ethylenedithiolate
MMe_3CH_2Ph	trimethyl-phenylmethylammonium(1+)
mpt	N-methylpiperidine-4-thiolate
NC_6H_{10}	hexylideniminate(1-)
$N(H)C_6H_4N(CMe_3)$	1-alkyl-2-aryldiazene
N-metlut	N-methyllutidinium(1+)
NP_3	$N(CH_2CH_2PPH_2)_3$
$(OBu^t)_2Si$	di-tert-butoxysilanediyl
OC_3Otpp	1,2,4,5-substituted benzene cap with five atoms areas of the type –$O(CH_2)_3O$- linking the benzene cap to the ortho positions of 5,10,15,20-tetraphenylporphyrin
$ONCMe_2$	2-proximate
pan	p-anisylphosphine
pc	phtalocyaninate(2-)
$P(p\text{-}C_6H_4Me)_2$	di(p-tolyl)phosphide
pcp	pentamethylcyclopentadienyl[phenyl(trimethylsilyl)methylene] phosphane
PC_2Pr^iPh	ligand prepared via reaction phosphide sulphides with $Fe_2(CO)_9$
Pcy_2	dicyclohexylphosphine
$PcyPh_2$	cyclohexyldiphenylphosphine
$P(2\text{-futyl})_3$	tri(2-furyl)phosphine
$PhCN_2S_2$	phenyl-1,2,3,5-dithiazolyl
Ph_2sic	4-neopentyl-2,3-diphenyl-1-silylenacyclobut-2-ene
pmpe	trans-2,2,3,4,4-pentamethyl-1-phenylphosphetane
$P(OC_6H_4Me\text{-}o)$	o-tolylphosphite
pp	2,3-bis(diphenylphosphino)-N-methylmaleic anhydride
$(PPh)_4$	cyclocatenate
PPh_2CH_2COO	diphenylphosphinacetate(1-)
$PPh(PPh_2)_2$	1,1,2,3,3-pentaphenyltriphosphane
ppn	bis(triphenylphosphine)iminium(1+)
$P_2(Pr^i{}_2NB)_2(SiPH_2)$	2,4-bis(diisopropylamino)-5-diphenyl-1,3-diphospha-2,4-dibora-5-silabicyclo[12.1.1]pentane
$(PrBNBu^t)_2$	1,3-di-tert-butyl-2,4-dipropyl-1,3,2,4-diazadiboretane
Pr^I-cma	tris(N-isopropylcarbamoylmethyl)amine
$Pr^i{}_2NPO$	diisopropylaminophosphoxane
PS3*	tris(3-phenyl-2-thiophenyl)phosphine
ptha	tetrafunctionalized p-tert-butylcalix[4]arene
py	pyridine
qt	quinoline-2-thiolate
$(SBu^t)_2Si$	bis(tert-butylthio)silanediyl
Sbz	benzenethiolate(1-)
SC_6H_9	cyclohexen-2-yl-thio

S_2CPcy_3	dithiocarboxy ligand
$S_2C_2(CF_3)_2$	cis-1,2-di(perfluoromethyl)-ethylene-1,2-dithiolate(2-)
S_2C_{60}	sulfido derivat of C_{60} fluorene
$S_2C_2(SMe)_2$	dimethyltetrathiooxalate
SC_6H_4-4-OH	4-hydroxythiophenol
$S_2H_8C_7$	1,3-dithiolatocyclohepta-4,6-diene
Sepy	pyridine-2-selenole
sil	[2-[(dimethylamino)methyl]-phenyl]phenylsilanedryl
$SiMes_2$	dimethylsilylene
$S(OMe)CH_2$	methoxymethanethiolate(1-)
tatc	2,3,7,8-tetraazatetracyclo[7.2.1.0.^{4,11}.0$^{6.10}$]dodeca-2,7-diene 231
tb	1,4-diaza-1,3-butadiene
tcgd	tricyclo[2.2.1]gent-5-en-2,3-ylendithiolate
$tmpBS_2$	tetramethylpiperidinoborano-dithiolate
tol	toluene
tpp	5,10,15,20-tetraphenylporphyrinate(1-)
tppm	tetraphenyldiphosphinomethane
tpp piv6	5,10,15,20-tetraphenylporphyrin macrocycle with two lateral pivalamido pickets (n=8)
η^1-tsamp	tetrasilytated amino(methylene)phosphine

1. INTRODUCTION

A characteristic feature of the d-group transitions metals is their ability to form complexes with a variety of neutral molecules such as carbon monooxide and iron is not exception. Very diverse types of complex exist. In many of these complexes the metal atoms are in low-positive, zero or negative formal oxidation states. It is a characteristic of the such ligand as CO which can stabilize low oxidation states, this property is associated with the fact that the CO group posses vacant π orbitals in addition to lone pairs.The most important π-acceptor ligand is carbon monoxide. Many of its complexes are of considerable structural interest as well as having importance industrially and in catalytic and other reactions.

Many structural studies of iron compounds have been carried out. To date there has been comprehensive overview of structural data of monomeric [1], dimeric and oligomeric coordination compounds [2]. Variations and trends in bond lengths and angles were found and discussed.

The systems discussed in this manuscript are iron carbonyls. The structures have been grouped by degree of nuclearity from monomers to oligomers.

2. MONONUCLEAR IRON CARBONYLS

2.1 Iron Carbonyls with Coordination Number Four

Crystallographic and structural data for four coordinated iron carbonyls are gathered in Table 1. There are six examples in which an iron atoms are a tetrahedrally coordinated with a different degree of distortion. Three derivatives [3,4] contain four carbonyls which created a tetrahedral environment about the iron (-II) atoms. In another two examples [5,6] complex anion $[Fe(CO)_3(NO)]^{-1}$ contain Fe(-II) atom surrounded by three carbonyls and one NO group, which build up a tetrahedral geometry about the Fe(-II) atom. In brown $[Fe(CO)(NO)_2(PPh_3)]$ [7] three different ligands, created a tetrahedral configuration about the iron atom FeN_2CP (Table 1). The mean Fe - L bond distance elongate with an increases covalent radius of coordinated atom in the order: 1.66Å (N, 0.75Å) < 1.745Å (C, 0.77Å) < 2.26Å (P, 1.Å). The L - Fe - L bond angles range from 101.0 to 129.7 (2)°, which indicates a distortion of tetrahedral geometry.

2.2 Iron Carbonyls with Coordination Number Five

The relevant data for mostly yellow, orange and colourless five coordinated iron carbonyl compounds are listed in Table 2. There are about one hundred such derivatives, which are listed and referenced in order of increasing covalent radius of the principle coordinated atom. Both of the two principle geometries, trigonal-bipyramidal and square-pyramidal, are represented here in varying degrees of distortion. The dominant type is trigonal-bipyramidal and iron in its oxidation states zero. Structure of ruby-red $Fe(CO)_3$(hpfb) [71] is show in Fig.1 as an example. The iron (0) atom is five-coordinate with approximately trigonal-bipyramidal geometry. The molecule has approximate non-crystallographic mirror symmetry. The Fe(0) atom coordination departs from ideal trigonal-bipyramidal symmetry, possibly caused by the close approach of the cyclohexane groups on two P atoms. The hpfb ligand with P(1) in the equatorial position and P(2) in the axial position, the resulting five-membered chelate contributes substantially to the stability of the complex.

The trigonal-bipyramidal environments around Fe(0) are built up by five unidentate ligands and three unidentate plus one bidentate set of ligands, the former dominates. The Fe(0) atoms in this series are found with the chromophores: FeC_5 [8 - 13], FeC_4H [14,15], FeC_4N [16 - 18], FeC_4P [19,22 - 46], FeC_4S [47], FeC_4Si [48 - 54], FeC_4As [26,55], FeC_4Sb [56], FeC_4Bi [57], FeC_3N_2 [58], FeC_3P_2 [33,37,59 - 62,64 - 68,70 - 76], FeC_3As_2 [77], FeP_3C_2 [78 - 80], FeC_3HP [81 - 83], FeC_3PSi [84,85], FeS_3CP [86], FeN_2P_2C [87], FeC_2S_2P [88,89], FeC_2P_2N [90 - 92], FeC_2P_2S [93,94], FeS_2P_2C [95], FeP_2HNC [96,97] and FeP_2HCSi [98].

There are examples, in which the iron (II) atom adopts a square-pyramidal environment, with the chromophore FeN_4C [20,21]. In one example [69] three C and two P donor atoms around the iron (0) atom created an intermediate between trigonal-bipyramidal and square-pyramidal arrangement.

As one expect, most common ligands is carbonyl, noticeable, the mean Fe(0) -C bond distance of 1.77Å is found for both equatorial and axial positions. The Fe(0) - C bond distance in an equatorial position range from 1.66 to 1.83Å, which is somewhat wider than those found in an axial position (range 1.68 - 1.81Å), respectively. In all other Fe(0) - L bond distances are generally shorter for equatorial than axial ones. For example, Fe(0) - L equatorial versus axial distances increases in the sequences: 2.05 vs 2.29Å (SL) < 2.17 vs 2.24Å (PL) for monodentate; and 2.16 vs 2.22Å (SL) < 2.02 vs 2.225Å (PL) < 2.31 vs 2.36Å (AsL) for bidentate ligands. In general, the mean Fe(0) - L bond distance increase as the covalent radius (in parenthesis) of, donor atom increases, in the orders: 1.59Å (H, 0.37Å) < 1.77Å (CO, 0.77Å) < 1.86Å (CL, 0.77Å) < 2.06Å (NL, 0.75Å) < 2.09Å (SL, 1.02Å) < 2.22Å (PL, 1.06Å) < 2.25Å (SiL, 1.17Å) < 2.38Å (AsL, 1.20Å) < 2.47Å (SbL, 1.40Å) < 2.67Å (BiL, 1.46Å) , for monodentate; and 2.18Å (SL, 1.02Å) < 2.21Å (PL, 1.06Å) < 2.335Å (AsL, 1.20Å) for bidentate ligands.

Both electronic and steric factors associated with the donor atom appear to influence the L - Fe - L bond angles of the metallocyclic rings. The L - Fe - L bond angle of the respective metallocyclic opening as the covalent radius of, donor atom decrease. The mean values for the five-membered angles are: 84.0° (As - donor) < 85.4° (P + S - donor) < 86.5° (P - donor) < 87.5° (S - donor).

Two isomeric forms, differing by degree of distortion, is observed for $[Fe(CO)_3(PPh_3)_2]$ - orthorhombic [61] and triclinic [62]. A pair of distortion isomers, within the same crystal is found for several examples: $[Fe(CO)_4\{P_2tmpB)_2(SiMe_2)\}]$[35], $[Fe(CO)_3(C_5H_{10}N_2)(PEt_3)]$[46], $[Fe(CO)_4\{Me_2Si(hmpa)\}$[52], $[Fe(CO)_2)(CS)(PPh_3)_2]$ [75] and $[FeH(CO)(NO)(PPh_3)_2]$ [96].Even three crystallographically independent molecules have been found in $[FeH_3(CO)(dppe)\{Si(OEt)_3\}$ [98]. The coexistence of two or more species, differing only by degree of distortion, is typical of the general class of distortion isomerism [99].

2.3 Iron Carbonyls with Coordination Number Six

Crystallographic and structural data for almost seventy six-coordinated iron carbonyls are gathered in Table 3. They are predominantly red, purple or amber while in the series of four-coordinated (tetrahedral) dominant iron in its oxidation state -2, in the five-coordinated (trigonal-bipyramidal) a zero; in the six-coordinated it is +2.

The structures are referencing in the order of increasing complexity of the coordination sphere and increasing covalent radius of the principal coordinated atom.

There are three principal forms of distortion of the octahedron. One is tetragonal, elongation or contraction along a single C_4 axis; the resultant symmetry is only D_{4h}.

Another is rhombic, changes in the lengths of two of the C_4 axis so that no two are equal; the symmetry is then only D_{2h}. The third is a trigonal distortion, elongation or contraction along one of the C_3 axis so that the symmetry is reduced to D_{3d}. All of these distortion are observed in the six-coordination chemistry of iron.

There is no example of a homogeneous chromophore around iron atom. Environment about iron atoms are created by six unidentate ligands [57,112,115,116,118,119, 124,125,129,132,134,139,140,142,143], four unidentate plus one bidentate [110, 111,117, 133,138,144], three unidentate plus one tridentate [113,123], two unidentate plus one tetradentate [20,100-108,121,122,126,130,131,141] and one unidentate plus one pentadentate [108,109,127,128].

Structure of a deep red [$Fe(CO)_3(\eta^3$-S_2CPcy_3) [113] is shows in Fig. 2 as an example. The molecule consists of a pyramidal $Fe(CO)_3$ fragment, to which tricyclo-hexylphosphonio dithiocarboxy ligand is bound in a pseudo allylic η^3 fashion through the two sulphur and the central carbon atom. The overall geometry of the molecule can be approximated to a C_S symmetry, with a staggered disposition of the pseudoallyl SCS group with respect to the three carbonyl ligands. The two Fe -C(carbonyl) distances irons to the sulphur atoms [1.76(1) and 1.78(1)Å] are shorter than the Fe - C(carbonyl) trans to the central carbon of S_2CPcy_3 [1.84(1)Å]. This respect a larger trans influence of the coordinated carbon with recpect to the suplhur atoms.

There is wide range of donor atoms which build up a distorted octahedral environment about iron with the chromophores: FeN_5C [20,100 - 107], FeS_5C [108,109], FeC_4X_2 (X = B [110 - 112]; S [113]; P [114]; Si [115,116] or Sb [117]), FeS_4C_2 [108, 119 - 122], FeN_3C_3 [123], FeC_3X_3 (X = S [124] or Te [125]), FeN_4OC [126], FeS_4OC [127], FeS_4NC [127,128], FeS_4CI [108], FeP_4HC [129], FeP_4CCl [130], FeN_3S_2C [131], FeC_3H_2As [132], FeC_3P_2O [133], FeC_3I_2Te [134], $FeO_2C_2P_2$ [135], $FeN_2C_2X_2$ (X = S [136] or Se [137], $FeC_2S_2P_2$ [138], $FeC_2P_2Se_2$ [139], FeC_3HPSi [140], FeP_3HNC [141], FeC_2P_2HS [142], FeC_2P_2HI [143] and FeC_2P_2OI [144]).

Interestingly, that the mean Fe(II) - C(carbonyl) bond distance of 1.77Å is the same as that found in the series of five-coordinated derivatives, where dominant is Fe(0) atom (Table 2).

Comparing other monodentate ligands, the Fe(II) - L bond distances increase in the order: 1.43Å (H) < 1.905Å (CL) < 2.07Å (OL) < 2.08Å (NL) < 2.235Å (PL) < 2.30Å (AsL) < 2.34Å (SL) < 2.36Å (Cl) < 2.41Å (SiL) < 2.45Å (SeL) < 2.61Å (I) ~ 2.61Å

(TeL) < 2.75Å (SbL). The Fe(II) - C(L) and Fe(II) - Si(L) bond distances are somewhat longer than those found in the series of five-coordinated iron(0) species, all remaining are shorter.

The multidentate ligands include both homo- and hetero-donor atom. The mean Fe(II) - L bond distances for homo-bidentate ligands increase in the sequence: 2.00Å (CL) < 2.20Å (BL) < 2.26Å (PL) < 2.29Å (SiL) < 2.30Å (SL) < 2.715Å (SbL) and for homo-tetradentate the order is: 1.94Å (NL) < 2.25Å (PL) < 2.28Å (SL). The tridentate ligands include those with one C plus two S and those with three N-donor sites. The hetero-tetradentates are either two N plus two S and one N plus three P-donor sites. The pentadentate ligands include one O plus four S, one N plus four S and five S-donor sites. In general, the mean Fe(II) - L bond distance shortened with an increase of dentaties of the respective ligands with the mean Fe(II) - N values: 2.08Å (mono-) < 2.025 Å(tri-) < 1.95Å (tetradentate); and for Fe(II) - S the values are: 2.34Å (mono-) < 2.30Å (bi-) < 2.28Å (tri-) < 2.26Å (tetradentate).

The effect of both eletronic and steric factors are observed in the values of the L - Fe - L bond angles of the metallocyclic rings. In three-membered rings the mean L - Fe - L intra-ligand angles opening in the order: 38.2º (O plus C-donor) < 47.5º (C plus S-donor) < 48.0º (B-donor) < 61.5º (Sb-donor). For the four-membered rings the values are: 70.5º (N plus S-donor) < 72.0º (N plus Se-donor) < 73.5º (C plus P-donor) < 78.5º (S-donor). For five-membered rings the values increase in the order: 84.5º (N-donor) < 85.0º (O plus S-donor) < 85.5º (P-donor) < 86.3º (N plus P-donor) < 87.8º (N plus S-donor) < 88.2º (S-donor). For six-membered rings the values are: 91.8º (N-donor) < 93.7º (N plus S-donor).

Two isomeric forms, cis- and trans- are observed for $[Fe(CO)_2)(PPh_3)_2(H)I]$ [143]. Two crystallographically independent molecules differing by degree of distortion have been found in several species [101b,113,127,133b,137 - 139]; their are further examples of distortion isomerism [99].

2.4 Iron carbonyls with coordination number seven and eight

There are only four derivatives, in which iron atoms are surrounded by seven donor atoms and their crystallographic and structural data are given in Table 4.

X - ray analysis of a brown derivative [145] shows that the derivative contains a PPh_4^+ cations and a $[Fe(As_3Se_3)_2(CO)]^{2-}$ anions. The dianion, contains two As_3Se_3 ligands each coordinated to the central Fe(II) atom by two arsenic atoms and one selenium atom, as can be see on Fig. 3. Each of the As_3Se_3 ligands chelates to the iron atom to form two corner-sharing birdcage fragments. The two rings are twisted and skewed, making for a very irregular coordination geometry around the iron center. There is also a single CO coordinated to the iron center, making it seven-coordinate.The mean value of the three-membered metallocyclic ring, As - Fe - As, is 58.3(1)°. There are relationships between the Fe - L bond distance, the covalent radius of donor atom and the metallocyclic ring. With an elongation of the former and an increases size of the donor atom, the three-membered metallocyclic ring opens. For example: B - atom: 2.20Å, 0.82Å and 47.9° (Table 3); As - atom: 2.44Å, 1.20Å and 58.3° (Table 4) and Sb - atom: 2.715Å, 1.40Å and 61.5° (Table 3).

Molecule of red complex [146] contains NBu_4^{n+} cations and $[Fe(CO)_3(B_5H_7)(H)]^-$ anions. The central iron(0) atom is surrounded by the three carbon atoms of CO grous, three boron atoms of B_5H_7 group and completed by H atom (Table 4).

In another derivative [147] one 1,3-di-tert-butyl-2,4-dipropyl-1,3,2,4-diazadiboretidine as tetra-hapto-N_2B_2 ligand with three carbonyl groups build up a piano - stool about the central iron(0) atom ($FeC_3N_2B_2$) (Fig. 4, Table 4).

In light yellow derivative [148] a piano - stool around the iron(II) atom is created by one tetra-hapto-(B_2P_2) ligand and by three carbonyl groups ($FeC_3B_2P_2$) (Table 4).

There are four (yellow (x3) and red (x1)) derivatives [149], where each iron(II) atom is eight coordinated. Structure of $Fe(C_2B_9H_{11})(CO)(MeCN)(PPh_3)$ is shown in Fig.5, as an example. The closo 12-vertex icosahedral geometry comprising a polyhedral FeC_2B_9 framework and the pseudooctahedral coordination exhibited by iron(II) atom are common structural features displayed by all four ferracarboranes. In addition, each displays monodentate ligands which completed an eight coordination around each iron(II) atom (Table 4).

The effect of both electronic and steric effects of the ligands can be seen in the opening of the L - Fe - L bond angles of the respective metallocyclic rings. For the three-membered rings the mean L - Fe(II) - L bond angles increases in the sequence: 45.5° (C-donor) < 46.8° (C + B -donor) < 49.2° (B-donor).

3. BINUCLEAR IRON CARBONYLS

The crystallographic and structural data for the binuclear carbonyl iron compounds are gathered in Table 5. The structures are arranged in the order of increasing

Fe - Fe distance. There are two hundred derivatives in which the most common oxidation number of iron is zero. Most of the derivatives are red and orange,with some brown, yellov, purple, green and blue.

There are several types of bridging, the distorted edge-shared bi-octahedral structure (Fe - Fe bond included) with two bridging atoms being the most common.

The crystal structure of orange $Fe_2(\mu\text{-}PhNSO_2NPh)(CO)_6$ [151] is shows in Fig.6 as an example. The two bridging N-atoms of $PhNSO_2NPh$ ligand bring the iron atoms to within 2.378(1)Å, which is one of the shortest distance found in the iron carbonyls dimers. The corresponding mean Fe - N - Fe angles are 72.7(1)°. Two nitrogen atoms also serve as bridges in another ten derivatives [150 - 154,156,159,160,163,171,177,183]. The Fe - Fe bond distance range from 2.372(2) to 2.502(2)Å and the Fe - N - Fe bridge bond angles from 72.5(7) to 81.1(2)°.

Howerer, the most common bridging involves two sulphur atoms [161 - 166,168 - 170,172,174 - 176,178 - 180,182,184 - 186,188,190,192,194,196,198,199,201 - 203] and two phosphorous atoms [167,210,207,217,219,223,226,227,233,237,242 - 244,247, 248,251 - 253,255,257,258,260,261,266,269,271,273,275,277,284,289,291,298]. In several other cases, two carbonyl groups are used as bridges [116,160,212,231,240,241, 249,256,264], two selenium atoms [187,191,197,200,204,213,220,290,292], two tellurium atoms [145,224,226,228,246,250,254], two silicon atoms [262,270,278], two arsenic atoms [274] and two antimony atoms [283]. There are also examples which contain two different atoms which serve as bridges between the iron atoms; for example oxygen plus sulphur [157], oxygen plus phosphorous [189, 207], nitrogen plus carbonyl group [160], nitrogen plus sulphur [155,158], nitrogen plus phosphorous [244], hydrogen plus carbonyl group [193], hydrogen plus sulphur [234]. carbonyl group plus sulphur [173], carbonyl group plus phosphorous [236,239], sulphur plus selenium [181], sulphur plus tellurium [181], sulphur plus phosphorous [251,218,225], sulphur plus arsenic [235], selenium plus tellurium [205,206], chlorine plus phosphorous [210,211], phophorous plus selenium [222], phophorous plus tellurium [222] and phosphorous plus iodine [229].

The second type of bridging involves three atoms [221,239,252,276,285,286, 295]. In some derivatives besides two single atoms serve as bridges bidentate mostly dppm molecules in a syn - syn arrangements.

There are derivatives]195,244,260,203,267,293,294] which involves a single atom as bridge. Another few derivatives [221,230,268,272,276] involves besides a single atom a bidentate ligands in a syn - syn arrangements.

There are also some derivatives [279 - 282] in which only direct iron - iron bond kept two iron polyhedrons.

The remaining derivatives (Table 5) are all structurally unique. X - ray analysis of brown $[(PPh_4)_2[Fe_2(CO)_4(PSe_5)_2]$ shows [290] that the cluster contains two iron dicarbonyl fragment bridged by selenium atom from two PSe_5 groups (Fig. 7). The Fe...Fe separation of over 3.6Å ruled out a direct metal - metal bond. The coordination environment around each iron is only slightly distorted from octahedral (Table 5).

Structure of red-orange $(PPh_4)_2[Fe_2(CO)_4(SbSe_4)_2]$ [292] consists of well separated dianions of $[Fe_2(CO)_4(SbSe_4)_2]^{2-}$ an $(PPh_4)^+$ cations with an inversion center located between the two iron atoms. The anoins are dimers of iron carbonyl fragments which are bridged by $SbSe_4^{3-}$ group. The $SbSe_4^{3-}$ groups are not tetrathiometallates as might be expected, but rather consists of trivalent attached $\{Se - Se - Sb(Se)Se\}^{3-}$. The Fe...Fe separation is 3.633(2)Å with the mean Fe - Se - Fe bridge angles of 96.5(1)°. Each iron atom is six-coordinated ($FeSe_4C_2$).

In another orange derivative [295], two $[Fe(CO)_3]^+$ fragments are double bridged by a $(Te_4)^{2-}$ unit and also bridged by an usual $TeCl_2$ moiety (Fig. 8). An unsual feature of this complex is the presence of both oxidized ($TeCl_2$) and Te_4^{2-} tellurium. The bridging $TeCl_2$ unit has a trigonal-bipyramidal arrangement of electron pairs with the lone pair occupiyng an equatorial site. Each iron atom attains an 18-electron configuration without formation of an Fe...Fe bond.

In another derivative [79] two $Fe(CO)_2\{P(OMe)_3\}_2$ fragments are held together by N_2 molecule in the manner of Fe - N - N - Fe. Each iron atom is in a trigonal-bipyramidal environment with the chromophore FeC_2P_2N (Table 5).

X - ray analysis of yellow derivative [297] shows that the two $Fe(CO)_4$ fragments are held together by $P(H)NBu^tP(H)NBu^t$. The $P_2N_2(Bu^t)_2$ ring is planar. Each iron atom adopt a distorted trigonal-bipyramidal geometry (FeC_4P).

In another yellow derivative [303] two $Fe(CO)_3$ fragments are held together by a P_4 -dentate ptba ligands (by two P atoms to each iron(0) atom). Each iron atom has a pseudo-trigonal-bipyramidal environment (FeC_3P_2). The Fe...Fe separation of 7.060(2)Å is the longest one found in this group.

Inspection of the data in Table 5 reveals that there are a variety of coordination geometries observed: tetrahedral, trigonal-bipyramidal and pseudo-octahedral. The ligand cover the range from mono- to tetradentate. In the derivatives, where four donor atoms are coordinated to the iron atom, the mean Fe - L bond distance elongated in the sequences: 1.69Å (NO) < 1.74Å (CO) < 2.15Å (NL) < 2.22Å (PL) < 2.23Å (P_2L) and 1.90Å (μ- CO) < 2.21Å (μ- PL) < 2.27Å (μ- SL) < 2.28Å (μ- S_2L).

In the series of derivatives, where five donor atoms are bounded to the iron atom, the mean Fe - L bond distance for monodentate ligands increasses in the sequence: 1.57Å (H) < 1.69Å (NO) < 1.78Å (CO) < 2.22Å (NL) < 2.23Å (PL). For monodentate bridged ligands the sequence is: 1.98Å (μ-CO) < 2.23Å (μ-PL) < 2.28Å (μ-SL) < 2.375Å (μ-SiL) < 2.38Å (μ-Br) ~ 2.38Å (μ-SeL) < 2.50Å (μ-TeL) < 2.605Å (μ-I) < 2.65Å (μ-SbL). For bidentate ligands the order is: 1.89Å (N_2) < 2.23Å (PL) < 2.25Å (SL) < 2.42Å (SiL). For bidentate bridged ligands the order is: 1.97Å (μ-NL) < 2.25Å (μ-SL) < 2.38Å (μ-SeL) < 2.54Å (μ-TeL).

In the series of derivatives, where six donor atoms are coordinated to the iron atom, the mean Fe - L bond distance for monodentate ligands increasses in the sequence : 1.66Å (H) < 1.78Å (CO) < 2.365Å (SiL). For single atom bridged ligands the sequence is: 1.63Å(μ-H) < 1.98Å (μ-CO) < 2.27Å (μ-PL) < 2.305Å (μ-SL) < 2.35Å (μ-SiL) < 2.62Å (μ-Te). For bi-, tri-, and tetradentate ligands-terminal vs bridged, the order is: 2.29 vs 2.30Å (S_2L) < 2.36 vs 2.35Å (P_2L) < 2.44 vs 2.47Å (Se_2L) < 2.45 vs 2.435Å (Se_3L) < 2.45 vs 2.435Å (Se_4L) < 2.61 vs 2.62Å (Te_3L).

From the given orders one can see, that the Fe - L bond distance is elongated with the covalent radius of the coordinated atom. In general, the Fe - L (terminal) bond distances are somewhat shorter than those of the Fe - L (bridge) bond distances.

The multidentate ligands include except homo- also hetero- donor atoms. Heterobidentate ligands include those with O plus N, O plus C, O plus S, O plus Si, N plus S, C plus S, C plus P, N plus P, Cl plus P, S plus Se, S plus As or Se plus Te-donor sites. Noticeable, unsually the bulker atom in the series of the hetero-donor ligands, serve as a bridge.

The effect of both electronic and steric factor are observed in the values of the Fe - L - Fe bond angles. The mean Fe - L - Fe bond angle close with an increase of the covalent radii of bridged atom in the respective group of the periodic system, in the sequences: 100.8° (H, 0.37Å) > 77.2° (Cl, 0.99Å) > 70.0° (Br, 1.14Å) > 59.4° (I, 1.33Å); 68.7° (O, 0.73Å) > 67.6° (S, 1.02Å) > 64.2° (Se, 1.16Å) > 61.6° (Te, 1.36Å); 77.0° (N, 0.75Å) > 73.0° (P, 1.06Å) > 67.0° (As, 1.20Å) > 64.0° (Sb, 1.40Å) and 83.4° (C, 0.77Å) > 70.4° (Si, 1.17Å).

There are tendencies of an elongation of Fe - Fe distance with an open of a Fe - L - Fe bridge angle. For example, the values of Fe - Fe distances and Fe - L - Fe bridge angles, for L = N - atom, are: 2.372(2)Å and 72.4(7)° [150] < 2.461(1)Å and 75.4(2)° [160] < 2.493(1)Å and 81.6(2)° [171]; for C - atom; 2.523(1)Å and 77.6(2)° [195] < 2.623(1)Å and 80.9(2)°[249] < 2.709(2)Å and 85.6(2)° [264} < 2.742(4)Å and 86.9°[276]; for S - atom: 2.465(4)Å and 66.1(1)° [161] < 2.533(1)Å and 69.4(1)° [201} < 2.717(4)Å and 75.6(1)° [267] < 3.062(4)Å and 83.2(1)° [285] < 3.420(1)Å and 95.2(2)° [288]; for P - atom: 2.574(2)Å and $69.6(1)^0$ [217] < 2.631(3)Å and 72.06(4)° [252] < 2.699(1)Å and 74.6(2)° [260] < 2.727(1)Å and 76.1(2)° [273] < 2.819(1)Å and 80.0(2)° [284] < 3.50(4)Å and 102.4(2)° [289] < 3.961(1)Å and 118.6(1)° [247] < 3.995(2)Å and 124.1(1)°[293]; for Se - atom: 2.597(2)Å and 63.5(1)° [187] < 2.575(2)Å and 66.1(1)° [220] < 3.633Å and 96.5(1)° [292]; and for Te - atom: 2.585(6)Å and 60.98(4)° [224] < 2.656(5)Å and 63.3(1)° [224] < 3.921(8)Å and 97.1(3)° [145].

The bridge angles and covalent radii of the bridging atoms are also related. If the Fe - Fe distance remains the almost the same, the Fe - L - Fe angle closses as the radius of the bridging atom increase. Some examples are: 2.523(1)Å and 77.6(3)° (C, 0.77Å), 2,524(1)Å and 67.7(1)° (S, 1.02Å) [196] 2.527(1)Å and 64.1(1)° (Se, 1.16Å) [197], 2.526(1)Å and 59.5(1)° (Te, 1.36Å) [181]; 2.573(1)Å and 81.4(1)° (C, 0.77Å [216], 2.574(2)Å and 69.6(1)° (P,1.06Å) [217], 2.575(2)Å and 66.1(1)° (Se, 1.16Å) [220], 2.585(6)Å and 60.98 (4)° (Te, 1.36Å) [224]; 2.709(2)Å and 85.6(2)° (C, 0.77Å) [264], 2.709(1)Å and 73.85(4)° (P, 1.06Å) [252], 2.705(1)Å and 71.2(1)° (Si, 1.17Å) [262]: 3.630(3)Å and 105.5(1)° (P, 1.06Å) [291], 3.633(1)Å and 96.5(1)° (Se, 1.16Å) [292].

The lemon $[Fe_2(CO)_6\{\mu\text{-}P(H)(CF_3)\}_2$ exists in two isomeric forms, monoclinic and triclinic [258].There are several examples, $[Fe_2(CO)_6\{\mu\text{-}N(Me)Ph\}\mu\text{-}\{OCN(Me)Ph\}]$ [160], $[Fe_2(CO)_6\{\mu\text{-}(Me_2NB)_2S_2\}$ [178], $[Fe_2(CO)_6\{\mu\text{-}ac\}\{\mu\text{-}SBu^t\}$ [230], $[Fe_2(CO)_6(\mu\text{-}SPh)(\mu\text{-}PPh_2)]$ [245], $(Kcrypt)_2[Fe_2(CO)_6(Te_3)]$ [246], $[Fe_2(CO)_5(\mu\text{-}Pcy_2)(\mu\text{-}PBu_2^t)(PHPh_2)$ [252], $[Fe_2(CO)_4(\mu\text{-}CO)\{\mu\text{-}(EtO)_2POP(OEt)_2\}$ [259], $[Fe_2(CO)_8(\mu\text{-}SiMes_2\}]$ [270], $[Fe_2(CO)_4(\mu\text{-}SO_2)(\mu\text{-}SMe)(PMe_3)_2].0.5Et_2O$ [286] and $[Fe_2(CO)_6(Te_4)(TeCl_2)$ [295], which contain two crystallographically independent molecules, differing only by degree distortion and they are another examples of distortion isomers. Brown black $[Fe_2(CO)_6(\mu\text{-}S_2Ph_4C_6)]$ contains three crystallographically independent molecules, again differ mostly by degree of distortion, with Fe - Fe bond distances of 2.521(4), 2.526(5) and 2.530(4)Å [194].

4. TRINUCLEAR IRON CARBONYLS

Crystallographic and structural data for trinuclear carbonyl iron compounds are gathered in Table 6. The structures are tabulated in order of an increasing of the mean Fe - Fe distance. There are several types of bridging. There are sixty nine derivatives which contain triangle of iron atoms (Table 6A).

X-ray analysis of red-orange derivative [304] contains two $NMe_3CH_2Ph^+$ cations and complex $[Fe_3(\mu_3\text{-}O)(CO)_9]^{2-}$ anion (Fig. 9). The cluster of idealized C_{3v} symmetry is made up of an almost perfectly equilateral triangle of iron atoms. Each iron atom is bonded to three terminal carbonyl group. A unique oxygen atom sits on the top of the Fe_3 triangle with mean Fe - O bond distance of 1.892(3)Å. The average Fe - Fe bond distance of 2.484(1)Å is the shortest found in the serie of trinuclear carbonyl iron compounds.On the other hand, the mean Fe - Fe bond distance of 2.774(2)Å found in the orange-brown $[Fe_3(\mu\text{-}H)_3(\mu\text{-}Bi)(CO)_9$ [357] is the longest one in this series.

There are only three examples [306,318,345], which contain triangle of iron atoms with no additional any bridge atom. In all remaining derivatives (Table 6A) one, two or even three atoms serve as bridges of triangle iron atoms. The Fe - Fe bond distance in the series range from 2.459(2) to 2.831Å with the mean value of 2.635Å.

There is interdependence between the Fe - L - Fe bridge angles and the covalent radii of the bridge atom, former closes with an increases covalent radius of bridge atom. For example, the mean Fe - L - Fe bridge angle and covalent radius (in parenthesis) of L are: 108° (H, 0.37Å) < 83.2° (N, 0.75Å) < 79.2° (C, 0.77Å) < 74.2° (S, 1.02Å) < 72.6° (P, 1.06Å) < 69.1° (Se, 1.16Å) < 62.3° (Te, 1.36Å).

As can be see in Table 6A, the iron atoms are coordinated by four donor atoms [304,306,307,415,319,320,322,324,325,332,336,343,345,348,349], by five atoms [305, 307,308,334,335,355], by three and four [306,318], by four and five [306,309 - 312,314, 316,317,320,326 - 328,330,331,333 - 342,344 - 347,354,356] by four and six [313,321], by five and six [350 - 353] and only by six donor atoms [357].

The mean Fe - L bond distances increases in the sequences: four coordinated: 1.79Å (CO) < 2.195Å (PL) for terminal ligands; 2.01Å (μ-CO) < 2.14Å (μ-SL) < 2.18Å (μ-PL); 1.91Å (μ_3-NL) < 2.12Å (μ_3-PL) < 2.19Å (μ_3-SL) < 2.32Å (μ_3-Se) < 2.49Å (μ_3-Te); five coordinated: 1.79Å (CO) < 2.205Å (PL); 1.66Å (μ-H) < 1.94Å (μ-NL) < 2.00Å (μ-CO) < 2.19Å (μ-SL) < 2.21Å (μ-PL) < 2.57Å (μ-Te); 1.91Å (μ_3-NL) < 1.98Å (μ_3-CO) < 2.17Å (μ_3-BL) ~ 2.17Å (μ_3-PL) < 2.195Å (μ_3-SL) < 2.35Å (μ_3-Se) < 2.64Å (μ_3-Te) and for six coordinated: 1.78Å (CO) < 1.97Å (NL) < 2.67Å (I); 1.59Å (μ-H) < 2.095Å (μ-CO) < 2.19Å (μ-PL) < 2.63Å (μ-Te); and 2.22Å (μ_3-PL) < 2.27Å(μ_3-CO) < 2.64Å (μ_3-Bi).

In general, the mean Fe – L (terminal) bond distances are shorter than those of Fe – L (bridge) bond distances. Noticeable, while the mean Fe – CO bond distance is almost constant the other Fe – L bond distances increases with coordination number.

There are examples, [$Fe_3(\mu\text{-}CO)_2(CO)_8\{P(OMe)_3\}_2$] [310], $(NEt_4)_2$[$Fe_3(\mu_3\text{-}Te)$-$(CO)_9$] [319], [$Fe_3(\mu\text{-}CO)$ $(\mu\text{-}CNCF_3)(CO)_{10}$] [331], [$Fe_3(\mu\text{-}CO)_2(CO)_8\{P(OMe)_3\}$ (Bu^tCN)] [337], [$Fe_3(\mu\text{-}CO)_2(CO)_9(PPh_3)$] [338] and (ppn)[$Fe_3(\mu\text{-}H)(\mu\text{-}SO_2)(CO)_9$] [344], which contains two crystallographically independent molecules, differ mostly by degree of distortion. The $[Fe_3(\mu\text{-}H)\{\mu_3\text{-}O(Et)PBu^t\}(CO)_9]$ [334] contains three such molecules.

Study of $[Fe_3(\mu\text{-}CO)_2(CO)_{10}]$ [328,329] at four different temperatures (110K, 160K,250K and 320K) shown, that on decreasing the temperature, the bridging ligands become progressively more symmetric, the bridged Fe – Fe bond shortets and the atomic displacement parameters show a congruent behaviour with temperature.

Crystallographic and structural data for forty mostly red derivatives, in which the three iron atoms form an approximately isosceles triangle with two "shorter" Fe – Fe bond distances (below 2.9Å) and remaining one over 3.0Å, are summarised in Table 6 B. The structures are refered in the order of an increasing of the mean value of two "short" Fe – Fe bond distances.

In dark purple $[Fe_3(\mu_3\text{-}NPh)_2(CO)_8\{P(OMe)_3\}]$ [358] the three iron atoms form an approximately isosceles triangle, with Fe - Fe 2.423(1) and 2.478(1)Å and Fe...Fe 3.079(1)Å. All carbonyl groups as well as $P(OMe)_3$ group are terminal. The NPh ligands are triply bridging with the mean Fe - N bond distances of 1.950(2)Å.

Structure of red brown $[Fe_3(\mu_3\text{-}NPh)(\mu_3\text{-}PBu^t)(CO)_9]$ [362] is shown in Fig.10 as an example of this group. The cluster consists of an isosceles triangle of iron atoms with two Fe - Fe bonds (2.560(2) and 2.569(3)Å) and with capping phoshinidene and imido ligands above and below the metal triangle, with the nitrogen atom being essentially equidistant from the three iron atoms (average 1.96Å). However, the phosphorus atom is slightly closer to Fe(1) and Fe(2) (average 2.208Å) than it is to Fe(3) (2.251(4)Å). The Fe(2) - Fe(1) - Fe(3) angle of 80.0(1)° is about 20.0° degrees longer than the 60.0°(mean) angle reported for triiron carbonyl clusters with three Fe - Fe bonds (Table 6A).

The Fe - Fe bond distance in this series ranging from 2.421(1) to 2.861(1)Å and nonbonding from 3.044(1) to 3.632(1)Å. But there are some derivatives for which the nonbonding values are not given.

In this series of triiron clustres (Table 6B) the mean Fe - L - Fe bridge angles closes with an increase covalent radii (in parenthesis) of bridging atom in the sequence: 79.5 and 108.0° (N, 0.75Å) > 75.5 and 107.0° (P, 1.06Å) > 69.0 and 101.0° (S, 1.02Å) > 68.5 and 97.0° (Se, 1.16Å) > 64.5 and 93.5° (Te, 1.36Å).

Iron atoms are surounding by four-, five- or six donor atoms, from which by farprevails five coordination. In general, the mean Fe - L bond distance elongate with coordination number in the orders: four-, five- and six coordination: 1.77, 1.78 and 1.81 Å (L = CO); 2.18, 2.27 and 2.275Å (μ-PL); 2.14, 2.21 and 2.31Å (μ_3-S); and 2.14, 2.21 and 2.25Å (μ_3-PL).

The mean Fe - L bond distances for five-coordinated derivatives increases in the orders: 1.78Å(CO) < 2.01Å (NL) < 2.19Å (PL); 1.70Å (μ-H) < 2.27Å (μ-SL) < 2.275Å (μ-PL) < 2.38Å (μ-AsL) < 2.44Å (μ-SeL) < 2.54Å (μ-TeL); and 1.94Å (μ_3-NL) < 2.21Å (μ_3-PL) < 2.25Å (μ_3-S) < 2.36Å (μ_3-Se) < 2.53Å (μ_3-Te).

There are five clusters, [$Fe_3(\mu_3\text{-}S)\{\mu_3\text{-}NC_6H_4\text{-}4\text{-}Me\}(CO)_9$] [158], [$Fe(\mu_3\text{-}S)(\mu_3\text{-}PPr^i)(CO)_8(CN_2H_5)$] [368], [$Fe_3(\mu_3\text{-}S)(\mu_3\text{-}PPh)(CO)_9$ [371], [$Fe_3(\mu_3\text{-}PPh)_2(CO)_9$] [374] and [$Fe_3(\mu_3\text{-}PPh)_2(NCMe)_2(CO)_7$] [376], which contain two crystallographically independent molecules differ mostly by degree of distortion and are another examples of distortion isomers.

There are ten mostly (red) triiron carbonyl clusters, which contain only one Fe - Fe bond and their structural parameters are given in the Table 6C. X-ray analysis of red [$Fe_3\{\mu_3$-$(cyhN)_2C\}(CO)_{10}$] [382] shows that the $Fe_2(CO)_6$ fragment has "saw horse" structure with each of the two N atoms of the coordinated carbodiimide molecule bridging its two iron atoms with Fe - Fe bond distance of 2.409(3)Å and the carbodiimide carbon atom acting as an axial carbene ligand to a distorted trigonal bipyramidal $Fe(CO)_4$ moiety.

In another red-orange triiron carbonyl cluster [383] the $Fe_2(CO)_6$ fragment is bridging by hydrogen atom and by carbonyl group with Fe - Fe bond distance of 2.580(3)Å. In additional, PMe group bridges all three iron atoms. There are two iron atoms, with the chromophore FeC_4HPFe' and the remaining third with the chromophore FeC_4P (Table 6C).

Structure of dark red [$Fe_3(\mu_3$-$Te)_2(CO)_9(PPh_3)$] [384] consists of an isosceles triangle of iron atoms tethered by capping μ_3-Te atoms. The basal $Fe_2(CO)_6$ unit contains a Fe - Fe bond (2.585(1)Å), while the apical $Fe(CO)_3(PPh_3)$ is octahedral, two coordination sites being occupied by bridging tellurium atoms. Remaining two iron - iron separations of 3.943(1)Å (average) ruled out direct metal - metal bond. Structure of another triiron carbonyl cluster, [$Fe_3(\mu_3$-$Te)_2(CO)_9I$] [325] is isostructural with that of [$Fe_3(\mu_3$-$Te)_2(CO)_9(PPh_3)$] [384].

Red [$Fe_3\{\mu$-$P(Ph)P(Ph)$-1,2-$C_6H_4P(Ph)\}(CO)_{10}$] [358,379] contains two crystallographically independent clusters, differ mostly by degree of distortion (Table 6C). The molecules contain a di-μ-phosphido diiron, as well as aproximal (phosphine) tetracarbonyl moiety.

Structure of orange $[Fe_3(\mu\text{-}CO)\{\mu_3\text{-}\eta^3\text{-}Me_2PCH_2P(Me)CH_2PMe_2\}(CO)_{10}]$ is shown in Fig.11. As can be see, one P - C - P donor set together with the two iron atoms Fe(2) and Fe(3) form a stable five-membered Fe_2P_2C ring with Fe(2) - Fe(3) bond distance of 2.705(1)Å. The third phosphorus atom P(1) binds an $Fe(CO)_4$ group.

Structure of orange $[Fe_3\{\mu\text{-}\eta^3\text{-}(Et_2C(O)PPP(NEt_2)_2\}(CO)_{10}]$ [386] contains one three-membered $Fe_2{}^{1,3}P$ and two four-membered $Fe_2{}^{1,3}P_2$ and $Fe^{1,3}P_3$ rings, with Fe - Fe bond distance of 2.7333(19)Å. One of the phosphorus atom binds an external $Fe(CO)_4$ moiety (Table 6C).

X-ray structure of deep orange $[Fe_3\{\mu_3\text{-}PPh\}\{\mu\text{-}P(Ph)\text{-}1,2\text{-}C_6H_4PPh\}(CO)_9]$ [379] has a highly symmetrical Fe_3P_3 core with one Fe - Fe bond distance of 2.8285(11)Å.

In the series of Fe_3 isosceles triangle with one Fe - Fe bond distance, the Fe - Fe bond distance range from 2.409(3) to 2.8285(11)Å (Table 6C).

There are fifteen triiron carbonyl cluster without Fe - Fe bond and their structural data are gathered in Table 6D.

Dark red $[Fe_3(\mu\text{-}SPh)_6(CO)_6].2\ Me_2CO$ [387] and black $[Fe_3(\mu\text{-}SPh)_6(CO)_6]$ [388] are isostructural. The former contains two crystallographically independent clusters, differ mostly by degree of distortion. While the acetone solvate cluster crystallizes in space group P-1, the no solvated cluster in space group $P2_1/n$. Both have almost the symmetry D_{3d} with a linear arrangement of the iron atoms in the manner: $(OC)_3Fe(\mu\text{-}SPh)_3Fe(\mu\text{-}SPh)_3Fe(CO)_3$. In these high-spin cluster iron(II) atoms are six-coordinated, the central iron atom with the chromophore FeS_6 and terminal iron atoms have FeS_3C_3 chromophores. The Fe...Fe separation of 3.18Å (average) ruled out direct metal-metal bond. The average of Fe - S - Fe bridge angle is 82.0°.

There are eight clusters of the composition $[Fe_3(\mu_3\text{-}E)(CO)_{12}]$, (E = As [389 - 391]; Sb [392 - 393], Te [319], Sn or Pb [394]), consist of a cental E (main-group atom) surrounded by three $Fe(CO)_4$ groups in a trigonal bipyramidal arrangement, with an average Fe - E - Fe angel of 115.5° (E = As or Sb); 109.5° (Te) and 120.0° (Sn or Pb). The iron atoms adopt the conventional trigonal-bipyramidal arrangement with the main group atoms occupying axial positions. The mean Fe - E bond distance elongated in the sequence: 2.46Å (E = As) < 2.55Å (Sn) < 2.58Å (Sb) < 2.62Å (Pb) < 2.64Å (Te).

Red $(ppn)_2[Fe_3(\mu_3\text{-}As)(CO)_{12}]$ [389, 390] contains two independent ppn cations in general positions and three independent $[Fe_3(\mu_3\text{-}As)(CO)_{12}]^{2-}$ anions with crystallographically-imposed C_3 symmetry. Another two clusters, $[Fe_3(\mu_3\text{-}E)(CO)_{12}]^{2-}$ (E = Sn or Pb) [394] each contain two crystallographically independent cluster anions (Table 6D).

X-ray structure of $[Fe_3(Se)_2(CO)_{10}]^{2+}$ [395] is shown in Fig. 12. Dication containing a picnic-basket like structure. The base of this basket is formed by two μ-η^1-, η^2-Se_2 units bridging Fe(1) and Fe(2). The two Se_2 units are capped by the Fe(3) atom forming the "handle" of the basket. The environment around the capping Fe(3) is nearly octahedral. In contrast, the environment of Fe(1) and Fe(2) in the base of the basket are considerably distorted from octahedral.

X-ray analysis of red brown triiron carbonyl cluster shows [396], that the two $Fe(CO)_4$ group are bridged by a $[R_2N\text{-}P(H)\text{-}\{Fe(CO)_4\}]$ PH unit. The salient feature of the structure is the difference between the P(1) - Fe(1) and P(1) - Fe(2) bond distances, 2.315(4) and 2.328(4)Å, respectively, which are essentially identical and the P(3) - Fe(3) bond length is 2.250(4)Å.

Structure of deep orange cluster [$Fe_3\{\mu$-$Et_2NC(O)PPP(NEt_2)_2\}(CO)_{12}$] [386] contains a bent phosphine unit (P(1) - P(3) - P(2) = 85.5(2)°) with two diethylamino group bonded to one of the terminal phosphorus atoms and $Fe(CO)_4$ units coordinated both to the center Fe(2) - P(3) = 2.278(2)Å and other terminal (Fe(3) - P(1) = 2.223(4)Å) phosporus atoms. In addition an $Fe(CO)_3$ unit bridges the terminal phosphorus atoms to form a four-membered FeP_3 ring. Most unusual, however, is bridging the Fe(1) - P(1) bond to form a four-membered FePOC ring. There are no iron-iron bonds in this structure.

Structure of orange [$Fe_3\{\mu_3$-η^3-$Me_2PCH_2P(Me)CH_2PMe_2\}(CO)_{12}$] [385] exhibits a staggered arrangement of the three $Fe(CO)_4$ fragments bound to the "stretched" P - C - P - C - P skeleton of ligand (Table 6D).

Inspection of the data in Table 6 reveals that there one hundred and thirty five triiron carbonyl cluster mostly red. In principal, from the number of direct iron-iron bonds, the triiron clusters are divided to the four groups: with three, two, one and no direct iron-iron bonds, from which by far prevails the former clusters (70) (Table 6A). The Fe - Fe bond distance of 2.409(3)Å [382] is shortest one found in the series of triiron carbonyl clusters. Most common environment about iron atoms, is a pseudo-octahedral including Fe - Fe bond. Besides, carbonyl ligands, which of course prevail, phosphorus donor atom (S) are commonly. There are several relationships which were discussed in each subsection.

5. TETRANUCLEAR IRON CARBONYLS

Crystallographic and structural data for sixty tetranuclear iron carbonyls are gathered in Table 7. The structures are very complex. There are sixteen derivatives [393, 397 - 409] which contain planar (tetrahedral) array of four iron atoms (Table 7A). The structures are listed in the order of an increases metal-metal bonds.

Black derivative [397] consists of a planar array of four iron atoms with a quadruply bridging nitrene ligand on each side of a plane. The mean Fe - Fe bond distances of 2.406(1)Å is the shortest found in the series tetranuclear iron carbonyls.

Tetrairon cluster of deep green [398] shows a high degree of planarity. There are two pairs of Fe - Fe bond distances, 2.486(3) and 2.541(3)Å, respectively.The ONEt ligand tetrabridges the Fe_4 cluster with both the N and the O atoms. The NEt ligand tetrabridges the Fe_4 cluster as well with the N atom nearly equidistant from the iron atoms (Table 7A).

In black derivative [399] the $[Fe_4(CO)_{13}]^{2-}$ anion of pseudo-threefold symmetry consists of four tetrahedrally oriented iron atoms in which apical $Fe(CO)_3$ group is symmetrically coordinated by only iron-iron bonds to a basal $Fe_3(CO)_9$ fragment containing three identical $Fe(CO)_3$ group located at the corners of an equilateral triangle and bonded to another by iron-iron bonds. The thirteenth carbonyl group triply bridges the three basal iron atoms.

Deep green derivative [400] consists of a planar array of four iron atoms with a quadruply bridging sulphido ligand on each side of the plane. The shortest Fe - Fe bond, 2.489(3)Å, contains a bridging carbonyl ligand. Semi-bridging carbonyl ligands bridge the two adjacent Fe - Fe bonds, 2.532(2)Å. The longest Fe - Fe bond, 2.605(2)Å, has no carbonyl bridge.

In dark red derivative [401] two tetrahedra of iron atoms each face capped by one carbonyl group are symmetry related and ordered; the third tetrahedron of iron atoms similarly capped is disordered, lying on the crystallographic center of symmetry. The Fe - Fe distances are: apical-basal 2.610(2)Å and 2.592(2)Å; basal-basal 2.545(2) and 2.556(2)Å. The shortening could be due to the presence of the triply bridging carbonyl group on the basal face.

Structure of brown derivative [402] is similar to that of dark red [401]. In another two dark green species [403] a planar array of Fe_4 is a quadruply bridged by sulphido and PPh ligand on each side of the plane (Table 7A).

Structure of brown $[Fe_4(CO)_8(py)_4]$ [404] consists of a triangulated parallelogram of iron atoms, two of which are coordinated, at opposite vertices, each of four carbonyl ligands and the other are each coordinated by two pyridine ligands. The Fe - Fe separations along the periphery are an average of 2.534(4)Å and of the shortest diagonal is 2.759(3)Å.

Cluster $[Fe_4(\mu_4\text{-}Se)_2(\mu\text{-}CO)(CO)_{10}]$ [405] consists of a trapezoidal arrangement of four metal atoms which are bridged on each side by a quadruply bridged Se ligands. Structure of another two tetranuclear clusters, black and red [406] are similar to that of [405]. Here a trapezoidal arrangement of Fe_4 are bridged on each side by quadruply bridged Ptol-p ligands.

Structure of black $[Fe_4(\mu_4\text{-}P^1Ph)(\mu\text{-}P^2Ph_2)(\mu_4\text{-}\eta^2\text{-}PhP^3P^4)(CO)_{10}]$ derivative [407] consists of a slightly distorted square containing four iron atoms, two of them being surrounded by two terminal CO groups and the other two by three terminal CO groups. To this square is coordinated a phosphanediyl PhP(1) acting as a four electron donor. The phosphorus atom P(2) is bonded to two phenyl group and to the two adjacent $Fe(CO)_2$ groups. The main feature of interest in the molecule involves the P(3)P(4) fragment. Each of the two phosphorus atoms is coordinated to an $Fe(CO)_2$ and an $Fe(CO)_3$ moeties. Furthemore the P(3) is linked to a phenyl group.

There are seven tetranuclear iron clusters [410 - 416] which contain a butterfly arrangement of iron atoms with some other donor atom situated in the center of the cluster. The structures are refered in the order of an increases mean Fe - Fe bond distance (Table 7B). In the three of them, black [410] and two brown clusters [411,412] a four-coordinate nitride atom forms a Fe_4N core with the mean Fe - Fe bond distance of 2.58Å and the average Fe - N - Fe bridge angles of 88.8(2)° and 177.8(2)°.

In a typical butterfly array of four iron atoms in another brown derivative [415] the boron atom is nearly centered between the wing-tips of the Fe_4 array with the bridging hydrogens located between the boron and wing-tip Fe atoms. The Fe - B - Fe angle is 162° and the boron lies ~ 0.3Å from the line joining the wing-tip iron atoms.

In another thirteen derivatives, two $RXFe_2(CO)_3$ fragments are joined by a μ_4-X spiron atom (Table 7C). In four derivatives [417 - 420] both X are sulphur atoms; another two red [421,214] contain μ-SR and μ_4-Se atoms; in dark red [422] and orange [423] both X are selenium atoms; in another two red derivatives [424] two nonequivalent fragments, $Fe_2(CO)_8$ and $(\mu\text{-}Cl)Fe_2(CO)_6$ are joined by a μ_4-P or μ_4-As and in orange [425] two $Fe_2(CO)_8$ fragments are joined by a μ_4-Si.

The Fe-Fe bond distance range from 2.52 to 2.79Å. The Fe - X - Fe bridge angles open in the sequences: 67.12^0 (X=Cl) < 67.5^0 (Se) < 68.4^0 (S) < 69.3^0 (As) < 72.5^0 (P) <73.1^0 (Si); and 131.2^0 (S) < 131.3^0 (P) < 132.6^0 (As) < 135.3^0 (Se).

There are four clusters [426 - 429] in which two $XFe_2(CO)_6$ fragments are joined by >X-X< bond (Table 7D). In deep green cluster [426] two $S_2Fe_2(CO)_6$ fragments are linked by an >S-S< bond in a centrosymmetric butterfly structure with Fe - Fe bond distance of 2.523(1) (av.) and Fe - S - Fe bridge angles of 67.8(1)Å (av.).

Black cluster [428] adopts a double "butterfly" conformation caused by an intramolecular Se - Se linkage between two $SeFe_2(CO)_6$ units with Fe - Fe bond distance of 2.555(6)Å and Fe - Se - Fe bridge angles of 64.7(2) and 108.1(2)^0(av.).

In dark red derivative [429] two butterly-shaped cluster cores Fe_2SAs are linked together by an As- As bond of axial type with the mean Fe - Fe bond distance of 2.591(1)Å and Fe - As - Fe bridge angle of 67.3^0.

There are twenty derivatives (Table 7E) which structures are complex (unique). Orange red blocks [430] contain two $Fe_2(CO)_6$ units double bridge by a C_2S_4 unit. The molecule sits on a crystallographic centre of symmetry at the mid-joind of the C - C bond. The Fe - Fe bond distance is 2.486(2)Å and Fe - S - Fe bridge angle is 65.7(1)^0 (av.)

In another cluster [431] two $Fe_2(\mu\text{-}SMe)(CO)_6$ moieties are bridged by $\mu_4\text{-}\eta^2$-$SCH_2C_6H_4CH_2S$-m (there are two Fe_2S_2 butterfly-shaped skeletons). A pseudosymmetric center lies in the central part of the benzene ring of bridge $CH_2C_6H_4CH_2$. The Fe - Fe bond distance is 2.514(1)Å and Fe - S - Fe bridge angle is 67.5^0 (av.).

The $[Fe_2(\mu\text{-}\eta^2\text{-}B_2H_4)(CO)_6]_2$ exists in two isomeric forms, orthorhombic (red orange) and monoclinic (dark red) [433]. Structure of the forms is shows in Fig.13. The molecule consists of two distorted tetrahedral B_2Fe_2 cages related at a noncrystallographic C_2 axis joined along on Fe - B edge such that the two Fe and two bridgehead atoms define a butterfly structure with a dihedral angle of $162.7(1)^0$. The Fe - Fe bond distances are 2.591(1)Å in the former and 2.578(2)Å in the latter.

In red derivative [434] the quadruply phosphours atom caps a $Fe_3(CO)_{10}$ group and is bonded to a $Fe(CO)_4$ fragment. In another red derivative [435a] the anion consist of a triangular $Fe_3(CO)_9(\mu\text{-}H)$ moiety capped by a phosphorus atom, which binds a dengling $Fe(CO)_4$ grup. In black derivative [435b] the antimony atom which is four coordinate caps a closed triangle of iron atoms, which also functioning as an electron pair donor to a fourth iron fragment.

There are two another anions, $[Fe_4(\mu\text{-}H)(\mu_4\text{-}Sb)(CO)_{13}]^{2-}$ and $[Fe_4(\mu\text{-}H)_2(\mu_4\text{-}Sb)(CO)_{13}]^-$ [393] in which the Sb atom caps a Fe_3 triangle to form a tetrahedron, and an external $Fe(CO)_4$ group is attached to it.

In green derivative [385] one P - C - P donor set of $Me_2PCH_2P(Me)CH_2PMe_2$ ligand forming a stable Fe_2P_2C five-membered ring. The remaining donor position of the ligand coordinates with an $Fe(CO)_4$ group. The Fe - Fe bond distance is 2.595(5)Å.

In red derivative[436] two $\mu_4\text{-}\eta^2\text{-}P(Me)P(Me)$ ligands double bridged two $Fe_2(CO)_6$ fragments. The Fe - Fe distance is 2.618(1)Å. and the mean Fe - P - Fe bridge angle is $72.0(1)^0$.

X-ray structure of dark brown $(PPh_4)_2[Fe_4(\mu_3\text{-}Te)_2(CO)_{14}]$ shows [348] that the central $Fe_2Te_2(CO)_6$ fragment is connected to two $Fe(CO)_4$ fragments through the two tellurium atoms. The Fe- Fe distance is 2.641(5)Å.

The anion $[Fe_4(\mu_4\text{-}Sb)(CO)_{16}]^-$ [437] has a highly tetrahedral core in which the four coordinate antimony atom is surrounded by two isolated $Fe(CO)_4$ units and an $Fe_2(CO)_8$ group. The Fe - Fe distance in the $Fe_2(CO)_8$ group is 2.720(1)Å,which falls in the range of the other Fe - Fe single bonds.

In red derivative [393] the antimony atom with the mean Fe - Sb - Fe angles of 109.5^0 is tetrahedrally bound to four $Fe(CO)_4$ units. All iron atoms have a trigonal-bipyramidal coordination geometry with the Sb atom occupying an axial position.

Dark purple derivative [325] $[\{Fe(CO)_4\}_4(\mu_4\text{-}Te)]^{2-}$ dianion with a central telluride bonded to four $Fe(CO)_4$ fragment. The environment about the central telluride is tetrahedral, with Fe - Te - Fe bond angles averaging $109.5(1)^0$.

The $[Fe_4(\mu_4\text{-}Se_2)_3(CO)_{12}]^{2+}$ dication in orange red derivative [395] is built up of two Fe_2Se_2 units connected by a linking Se_2 unit. All of the iron centres are roughly octahedral (FeC_3Se_3) but there is the pinching effect of the sidebound η^2-Se_2 unit to each iron center. The Se - Te - Se angle of $56.72(2)^0$ is typical of all the Fe- Se environments.

Structure of dark red derivative [439] is shown in Fig. 14. The structure is cubane-like with the iron centres octahedrally coordinated (FeC_3As_3) and Fe- As - Fe bond angles of $102.6(1)^0$.

Structure of another red derivative [440] consists of two Fe_2Se_2 butterfly tetrahedra connected by a s-trans-1,3-butadiene moiety. The bridging butadiene group is attached both the Se atoms of each $(CO)_6Fe_2(\mu\text{-}Se)_2$ group such that the molecule is centrosymmetric with a crystallographic inversion centre at midpoint of C - C bond. In purple derivative [191] both Se atoms in each Fe_2Se_2 butterfly are attached to phenyl-acetylene, which forms a bridge between the two $(CO)_6Fe_2Se_2$ units.

Structure of violet derivative [441] is shows in Fig. 15. The structure consisting of two Fe_2Se_2 butterfly units attached to each other by means of a phenylacethylene molecule, which is bonded to all the Se atoms. A dppe group is attached to one Fe atom of each Fe_2Se_2 unit. The Fe atoms associated with the dppe group have two terminal carbonyl groups bonded to them each while the remaining two Fe atoms have three terminal groups bonded to each of them. The average Fe - Fe bond distance is 2.529(2)Å. and Fe - Se - Fe bridge angles are 63.8(1)^0. (Table 7E).

In summary, the over sixty tetranuclear iron carbonyls illustrate the rich variety of behaviours in the chemistry of iron. The iron centres show wide range of coordinated donor atoms, ranging from two to seven, from which five by far prevails. The overall structures can be described as planar/tetrahedral array of Fe_4 and two $RXF_2(CO)_3$ fragments joined by a μ_4-X spiran atom.

In the series four-coordinate examples, the maen Fe - L distance increases in the order: 1. 76Å (CO)< 1.82Å (μ-CO) < 1.825Å (μ_4-N) < 2.03Å (μ_3-CO) < 2.17Å (μ_4-P) < 2.27Å (μ_4-PL) < 2.32Å (μ_4-S) < 2.46Å (μ_4 -Sb). In the series of five-coordinated examples the orders are: 1.65Å (μ-H) < 1.895Å (μ-CO)< 2.205Å (μ-PL) <2.27Å (μ-SL) < 2.305Å (μ-CL) < 2.40Å (μ-Se); and 1.95Å (μ_4-OL) < 1.98Å (μ_4-NL) < 2.24Å (μ_4-BL) <2.245Å (μ_4-S_2) < 2.26Å (μ_4 -S) < 2.265Å (μ_4-PL) < 2.28Å (μ_4-P) < 2.34Å (μ_4-As_2) < 2.35Å (μ_4- Si) < 2.36Å (μ_4-As) < 2.38Å (μ_4-Se) < 2.53Å (μ_4-Sb) < 2.64Å (μ_4-Te). In the series of six-coordinate examples the orders are: 1.68Å (μ-H) < 1.97Å (μ-CO); and 1.94Å (μ_4-OL) < 2.02Å (μ_4-NL) < 2.335Å (μ_4-PL) < 2.41Å (μ_4-S) < 2.49Å(μ_4-Sb).

The mean Fe - L - Fe bridge angle opens in the sequences: 65.4^0 (L=Sb) < 65.7^0 (Se) < 67.0^0 (Cl) < 68.2^0 (S) < 73.0^0 (P) ~ 73.0^0(Si) < 79.0^0 (C) < 82.5^0 (B) < 88.5^0 (N); 104.2^0(Se) < 108.7^0 (P) < 109.5^0 (C) < 117.0^0 (Sb); and 130.4^0 (P) < 131.2^0 (S) < 134.3^0 (Se) < 139.8^0(Sb) < 177.5^0 (N).

The $[Fe_2(\mu\text{-}\eta^2\text{-}B_2H_4)(CO)_6]_2$ exists in two isomeric forms, orthorhombic and monoclinic [433], which differ mostly by degree of distortion. There are four derivatives, $[Fe_2(CO)_4(py)_2]_2$ [404], $Fe_4(\mu_6\text{-}\eta^3\text{-}PPPhPPh)(\mu\text{-}PPh_2)(CO)_{12}$ [407], $Fe_4(\mu_4\text{-}As)(CO)_{14}Cl$ [427] and $\{Fe_2(\mu\text{-}SPh)(CO)_6\}_2(\mu_4\text{-}S_2)$ [427] which contain two crystallographically independent molecules, differ again mostly by degree of distortion and are another examples of distortion isomers.

The shortest Fe - Fe bond distance, found in the series of tetranuclear iron carbonyls is 2.406(1)Å [397].

6. PENTANUCLEAR IRON CARBONYLS

Crystallographic and structural data four sixteen black pentanuclear iron carbonyls are gathered in Table 8. The structures are complex but the Fe_5 skeleton with a distorted square-based pyramidal geometry is most common.

The Fe_5 skeleton in $[Fe_5(\mu_5\text{-}N)(\mu\text{-}CO)_2(CO)_{12}]^-$ [442] has a slightly distorted square-based pyramidal geometry with the nitride nitrogen atom located near the basal plane. There are twelve terminal carbonyl groups; three on the apical iron atom, Fe(1), three on the unbridged basal iron, Fe(2), and two each on the remaining iron atoms Fe(3), Fe(3*) and Fe(4). The other two carbonyls bridge the Fe(3) - Fe(4) and Fe(3*) - Fe(4) interactions. The cluster possesses of symmetry, passing through Fe(1), Fe(2) and Fe(4), which causes Fe(3) and Fe(3*) to be in symmetry - related mirror - image positions. The range of Fe(apical) - Fe(basal) bond distances is 2.5635(8) - 2.5889(5)Å (2.5789Å mean). Distances between Fe(basal) atoms depend in the types of $Fe(CO)_n$ site involved. The unbriged Fe(2) - Fe(3) distance is 2.6389(5)Å and the bridged Fe(3) - Fe(4) distance is 2.5753(5)Å.

Structure of $[N(PH_3)_2][Fe_5(\mu_5\text{-}N)(\mu\text{-}CO)_2(CO)_{12}]$ [443] did not contain a plane of symmetry and each iron sites had a different arrangement of carbonyl groups. The nitride nitrogen atom is located at an average distance of 0.11(1)Å below the mean plane containing the four basal iron atoms. The range of Fe(apical) - Fe(basal) bond lenghts [2.563(3) - 2.611(3)Å] is somewhat longer than in [442]. Again distances between Fe(basal) atoms depend on the types of $Fe(CO)_n$ sites involved. The unbridged Fe - Fe diastance are 2.629(3)Å and 2.646(3)Å, while the bridged are 2.550(3) and 2.559(3)Å.

Structure of $Fe_5(\mu\text{-}H)(\mu_5N)(\mu\text{-}CO)_3(CO)_{11}$ [412] established a square-pyramidal arrangement of iron atom, Fe - Fe 2.57(2)Å average, with the nitride nitrogen atom located near the basal plane. Actually, the N atom is 0.093(2)Å below the basal plane away from the apical iron atom.

There are four derivatives [305,307b,443] which established a square-pyramidal arrangement of iron atoms, with the carbon atom located near the basal plane (Table 8). The mean Fe - Fe and Fe - μ_5C bond distances of 2.62Å and 1.91Å are longer than those found in derivatives which contain nitrogen atom near the basal plane, with the values of (Fe - Fe) 2.60Å and 1.88Å (Fe - μ_5N), respectively.

Another three derivatives which contain $[Fe_5(\mu_3\text{-}X)_2(CO)14]^{2-}$ dianion (X= S or Se [322], S [444]) are isostructural. The dianion contains two verter - fused tetrahedra oriented such that the capping sulfide (selenide) vertices of each tetrahedron are rotated.

Both, the mean Fe - Fe as well as Fe - S bond lengths of 2.60 and 2.19Å are smaller than that found in the selenide derivative, with the values of 2.62 and 2.315Å, respectively (Table 8).

The $(NBu_2)_2[Fe_5(\mu_3\text{-}S)_2(CO)_{14}]$ [322] contains two crystallographically independent molecules, differ mostly by degree of distortion, and it is another example of distortion isomerism.

Structure of $[Fe_5(\mu_3\text{-}S)_4(CO)_{12}]^{2-}$ [445] is shown in Fig. 16. The structure consists of a central Fe(3) atom coordinated by four S atoms from the two $Fe_2S_2(CO)_6$ fragments. The geometry about the central Fe(3) is severely distorted from tetrahedral with S - Fe - S "bite" angles (due to the $Fe_2S_2(CO)_6$ fragments) of $82.0(2)^0$ (average); the maining S - Fe - S angle are grouped as two smaller (average $107.6(2)^0$) and two larger (average 147^0) angles, with an overall average value of 127.3^0. All remaining four iron atoms has the chromophore FeC_3S_2. The Fe - Fe bond distances of 2.483 and 2.500Å are shortest found in the series of pentanuclear iron carbonyls (Table 8).

Structure of $Fe_5(\mu_8\text{-}\eta^4\text{-}N_2C_{11}H_{16}N_2)(CO)_{13}$ [446] is shown in Fig. 17. As can be see the central iron atom is coordinated by all four nitrogen atoms of a syn-bis-diazene ligand with Fe - N bond distance of 1.913(9)Å (average). In additional the central iron atom is coordinated by one terminal carbonyl group. The remaining four iron atoms are tetrahedrally coordinated (FeC_3N) (Table 8).

There are two derivatives, which contain $[Fe_5(\mu_3\text{-}X)_2(CO)_{17}]^{2-}$ dianion, (X=As [390] or Sb [436]). The anions are isostructural, consisting of a slightly distorted square pyramidal core assembled from two arsenic [390] or two antimony [436] atoms and three $Fe(CO)_3$ fragments. Each X atom also binds an external $Fe(CO)_4$ group. The base of the pyramid [390] is distorted by a displacement of the arsenic atoms out of the basal plane toward the apical iron atom (Fe(2)) and a deviation of the internal basal angles from the ideal value of 90^0. The basal As - Fe - As and Fe - As - Fe angles average 77.48(5) and $102.1(5)^0$, respectively. Also the $Fe(CO)_4$ groups are bent away from the apex iron slightly.

The Sb_2Fe_3 square prism [436] is also not an ideal prism, since the basal Fe - Sb - Fe angle (78.86 and 79.12^0). Also the Fe(2) - Fe)4) - Fe(3) capping angles is 86.75^0, whereas the Sb(1) - Fe(4) - Sb(2) capping angle is 78.45^0. The external irons, Fe(1) and Fe(5) lie 0.2356 and 0.1539Å respectively, below the plane of the square Sb_2Fe_2 base of the prism. The atoms of the Sb_2Fe_2 square have a mean deviation from the plane of 0.0462Å.

Inspection of the data in Table 8 reveals that iron atoms are surrounded by three-, four-, and five - donor atom of ligands, from which the four coordination is prevails. The mean Fe - L bond distance for four - coordinated elongated in the order: 1.77Å (L=CO) < 1.885Å (μ_5-N) < 1.92Å (μ_5-C) < 1.965Å (μ-CO) < 2.22Å (μ_3-S) < 2.315Å (μ_3-Se). For five-coordinated in the sequence: 1.79 Å (CO) < 1.875Å (μ_5-N) ~ 1.875Å (μ_5-C) < 2.31Å (μ_3-S) < 2.38Å (μ_3-As) < 2.525Å (μ_3-Sb). The Fe - L - Fe bridge angle opens in the order: 68.0^0 (L=Sb) < 69.0^0 (S) < 69.5^0 (Se) < 70.0^0 (As) < 87.5^0 (N) < 88.0^0 (C). The shortest Fe - Fe bond distance is 2.483Å [445].

7. HEXANUCLEAR IRON CARBONYLS

There are seventeen hexanuclear iron carbonyls and their crystallographic and structural data are summarized in Table 9. Crystals of black $(PPh_4)_2[Fe_6(\mu_6\text{-}N)(H)(\mu\text{-}CO)_3(CO)_{12}]$ [447] consists of discrete phosphonium cations and cluster anions. The cluster adopts an octahedral metal cage, with an intersticial nitride ligand. Under the D_3 idealize symmetry, all iron vertices are equivalent, being bound to one edge - bridging and two terminal carbonyls. The Fe - Fe bond distance range from 2.533(1) to 2.707(1)Å.

The molecular structure of $[Fe_6(\mu_6\text{-}N)(\mu\text{-}CO)_3(CO)_{12}]^{3-}$ [447] is strinkingly similar to that of the parent dianion. The Fe - Fe bond distance range from 2.558(2) to 2.685(2)Å.

The anion $[Fe_6(\mu_6\text{-}C)(\mu\text{-}CO)_4(CO)_9(NO)_2]^{2-}$ has a crystallographically twofold symmetry. There is no distortion of the iron octahedron to D_3 symmetry with an interstitial carbide ligand and with Fe - Fe bond distances from 2.590(2) to 2.720(2)Å and Fe - Fe - Fe bond angles $60.00(7) \pm 3.2^0$ and $90.00(7) \pm 1.8^0$.

The octahedral array of iron atoms with an interstitial carbide ligand is also present in $[Fe_6(\mu_6\text{-}C)(\mu\text{-}CO)(\mu\text{-}SO_2)(CO)_{14}]^{2-}$ [305]. The Fe - Fe bond distance range from 2.582(2) to 2.723(2)Å. There is an edge-bridging SO_2, two semibridging carbonyls, and one symmetrically bridging carbonyl. The central carbide atom is equidistant from all six iron atoms.

Another dianion, $[Fe_6(\mu_5\text{-}Sb)_2(CO)_{20}]^{2-}$ [437] has a distorted octahedral trans Sb_2Fe_4 core in which each Sb atom donates its external lone pair of electrons to an $Fe(CO)_4$ unit. The Sb - Fe bond's within the octahedral core cover a range from 2.517(2) to 2.703(2)Å. The eight Sb - Fe bonds within the octahedron fall into two sets; 2.672Å (average) are much longer than the other four Sb - Fe distances of 2.521Å (average). These distances alternate such that each iron has one long and one short Sb - Fe bond. The result is that the Fe_4 square of the octahedron is very puckered. Two external iron atoms are trigonal - bipyramidally surrounded by four carbonyls and an Sb atom which occupies an apical position.

Structure of deep purple, $Fe_6(\mu_4\text{-}P)_2(\mu\text{-}PMe)_2(CO)_{18}$ [448] is shown on Fig. 18. The two cluster units are linked by a P - P bond. A noncrystallographic two - fold axis perpendicular to the P- P bond relates the two clusters. The orientation adopted by the two cluster units about the P - P bond (Fe(6) .. P(3) ... P(2) ... Fe(3) = 41.6^0) is likely dictated by the packing requirements of the carbonyls. The mean Fe -Fe bond distance within the clusters is 2.714(4)Å and between them 3.591(4)Å.

X-ray analysis of dark green cluster shows [449] the centrosymmetric molecule to contain two (necessarily) identical Fe_3-triangles linked by the $Ph_2P(CH_2)_6PPh_2$ ligand. There are two CO ligands bridging one Fe - Fe bond and nine terminal CO groups. Each P atom of the dpph ligand is attached to one of the bridged Fe atoms and is trans to an unbridged Fe - Fe bond. The bridged Fe - Fe bond of 2.593(1)Å is shorter than that of the unbridged with the mean Fe - Fe distance of 2.691(2)Å (Table 9).

There are four clusters [322,444,445,450] which contain $[Fe_6(\mu_3\text{-}S)_4(\mu\text{-}S)_2(CO)_{12}]^{2-}$ dianion. The molecule consists of a rhombic planar Fe_2S_2 core which has both iron vertices chelated by $Fe_2(CO)_6S_2$ butterflies. The Fe - Fe axes of the butterflies are perpen-dicular to the plane of the core, creating a pseudooctahedral coordination enviroment around the core atoms. The Fe - Fe distances in the butterflies are 2.503Å (average). The Fe - Fe distance across the central rhombus is somewhat longer at 2.683Å (average). The mean Fe - S bond distances are 2.203Å (μ-S) and 2.305Å (μ_3-S).

There are two dark purple clusters [322,451] which contain $[Fe_6(\mu_3\text{-}Se)_4(\mu\text{-}Se)_2(CO)_{12}]^{2-}$ dianion, which are isostructural with the parent sulfide clusters. The Fe - Fe distances in the butterflies are 2.550(1)Å (average) and across the central rombus is 2.753(4)Å (average). The Fe - Fe bond distances are slightly elongated relatively to the sulfide, as expected to accommodate the longer capping selenide with the mean Fe - Se bond distances of 2.322(2)Å (μ-Se) and 2.419(3)Å (μ_3-Se).

Structure of dark red $[Fe_6(\mu_4\text{-}S)_2(\mu\text{-}CNEt_2)_2(CO)_{12}].0.5CH_2Cl_2$ [452], consists of three metal - metal bonded $Fe_2(CO)_6$ units linearly linked in butterfly configuration by two bridging suphur atoms, and terminated on both sides by bridging diethyliminocarbene ($CNEt_2$) ligands. The Fe -Fe distances are nearly equal, with the central metal - metal bond (2.544(3)Å) slightly longer than those which are also carbene - bridged (both 2.479(3)Å).

Structure of green black $\{[Fe_2(CO)_8(\mu_4\text{-}As)_2]_2[Fe_2(CO)_6]\}$ [453] consists of four $AsFe_2$ triangles linked by spiro-As atoms and an Fe -Fe edge. The mean Fe - Fe bond distances is 2.72(1)Å and the mean Fe - As - Fe bridge angle is $69.7(3)^0$ (Table 9).

Structure of dark red $Fe_6(\mu\text{-}S)_2(\mu\text{-}SC_6H_9)_2(CO)_{18}$ [454] is shown in Fig. 19. The cluster consists of three $Fe_2(CO)_6$ fragments joined to the two doubly bridging atoms S(2) and S(3) and that each $Fe_2(CO)_6$ grouping has an Fe - Fe single bond and two μ_4-S or one μ_4-S and one μ-SC_6H_9 bridge. The coordination of the iron atoms is distorted tetragonal-pyramidal (without Fe - Fe bond) (Table 9).

Inspection of the data in Table 9 reveals that in the seventeen hexanuclear iron carbonyls, iron atoms are surrounded only by four or five donor atoms. The mean Fe - L bond distance elongated in the sequences: 1.82Å (CO) < 1.86Å (μ_6-N) < 1.885Å (μ_6-C) < 1.935Å (μ-CO) < 2.18Å (μ-SL) < 2.23Å (μ-S) < 2.30Å (μ_3-S) < 2.32Å (μ-Se) < 2.41Å (μ_3-Se) for four coordinated; and 1.775Å (CO) < 1.885Å (μ_6-C) < 2.125Å (μ-CO) < 2.21Å (μ_3-PL) < 2.24Å (μ_4-P) < 2.26Å (μ_4-S) < 2.31Å (μ_3-S) < 2.375Å(μ_4-As) < 2.42Å (μ_3- Se) < 2.58Å (μ_4-Sb) for five coordinated.

There is no cluster without metal - metal bond(s) with the shortest in the series 2.479(3)Å [452]. The mean Fe - L - Fe bridge angles open in the order: 66.0^0 (L=Sb) < 68.2^0 (Se) < 70.0^0 (As) < 72.0^0 (S) < 81.5^0 (C).

8. HEPTA - AND OCTANUCLEAR IRON CARBONYLS

Black heptanuclear iron carbonyl of the composition $(AsPh_4)_2[Fe_7(\mu_7\text{-}B)(\mu\text{-}CO)(CO)_{19}(H)]$ [455] it is only example and structure of the dianion is shown in Fig. 20. Six of the iron atoms are arranged in an approximately trigonal prismatic geometry and encapsulate a single bond atom. The seventh iron atom, Fe(5), with two terminal carbonyl ligands caps a nearly square face with two Fe - Fe and two Fe - (μ-CO) - Fe interactions. Bridged interactions are shorter than unbridged (2.563(2) vs 2.635(2)Å), resulting in a slight fold along a line connecting Fe(1) and Fe(3). Both of these average are shorter than the Fe - Fe distances (average 2.74(3)Å) of the square face capped by Fe(5). The rest of the CO's are bound in terminal arrangements, however, the C(10)- and C(4)- metal distances suggest some semibridging character. The other two four- sided faces of the trigonal prism are rectangular, as expected. Crystallographic and structural data are given in Table 10.

There are three octanuclear iron carbonyls and their crystallographic and structural data are summarized in Table 10. Black derivative [428] contains two Fe_4Te_4 cubane - like bridged by a Te_2^{2-} Fig. 21. The latter acts overall as a μ_4-type ligand. These cages can be each described as being formed by the interpenetrating tetrahedral arrays of four iron and tellurium atoms. There are two types of iron atoms, four with the chromophore FeC_3Te_3 and the other four FeC_2Te_3. The mean Fe - Fe bond distance is 2.619(1)Å.

Dark red cluster [224,456] contains well separated $(NMe_3CH_2Ph)^+$ cations and $[Fe_8(\mu_3\text{-}Te)_6(CO)_{24}]^{2-}$ dianion. The dianion can be viewed as two $Te_2Fe_2(CO)_9$ fragments asymmetrically attached each to the two Te atoms of the $Te_2Fe_2(CO)_6^{2-}$ dianion.The dianion is located on the crystallographic 2-fold axis which passes through the centers of Fe(1) - Fe(1') and Te(1) - Te(1'), respectively. There are six iron atoms with the chromophore FeC_3Te_2 and remaining two $Fe_2C_3Te_3$. The mean Fe - μ_3Te bond distance in the former of 2.566(3)Å is about 0.07Å shorter than that found in the latter (2.636(3) Å). The mean Fe- Fe bond distance is 2.601(3)Å. (Table 10).

Structure of $[\{Fe_4(\mu_4\text{-}PPh)_2(CO)_{10}\}_2\{\mu\text{-}(MeO)_2PC_6H_4P(OMe)_2\}]$ [457] is shown in Fig. 22. As can be see, two equivalent $Fe_4(\mu_4\text{-}PPh)_2(CO)_{10}$ clusters are attached by $(MeO)_2PC_6H_4P(OMe)_2$. The double-cluster molecule has a crystallographic center of symmetry in the center of the aromatic ring. There are two types of Fe - Fe lenghts, with the mean values of 2.535(2) and 2.727(1)Å, respectively.

9. CONCLUSIONS

This rewiev has classified over five hundred and sixty iron carbonyls, which were characterized by X-ray diffraction. The various geometries formed are found in increasing number of order 8 < 7 < 6 < 4 < 5. The ligands range from mono- to octadentate. The nuclearity of the derivatives range from mono- to octanuclear, bi- and mononuclear examples by far prevail. Pairs of independent molecules with differing degree of distortion are formed in the same crystal in several cases [35,46,52,75,96,158,160,178, 230,245,246,252,259,270,286,295,310,319,322,331,337,338,344,368, 371, 374, 376, 378, 379,387,394,404,407,424,427]. In four cases [98,194,334,390] three such molecules are presented. Three derivatives, exist in two isomeric forms; orthorhombic [61] and triclinic [62]; monoclinic and triclinic [258], orthorhombic and monoclinic [433]. All of these belong to the general class of distortion isomerism [99].

Relationships between the various structural parameters have been discussed within each section. The shortest Fe - Fe bond distances in the series of binuclear species is 2.378(1)Å [150], in trinuclear is 2.459(2)Å [305], in tetranuclear is 2.391(1)Å [397]; in pentanuclear is 2.483Å [445], in hexanuclear is 2.479(3)Å [452], in heptanuclear is 2.560(2)Å and in octanuclear is 2.518(2)Å [457].

This review, together with its precusors dealing with over nine hundred monomer and over four hundred di- and oligomeric coordination compounds of iron [1,2] give overview of iron structural chemistry of iron carbonyls and coordination compounds. Previous reviews on cobalt [458] and nickel [459,450] complete, the overview of this triade. Despite the availability of increasingly sophisticated data retrieval systems, the area of crystallography remains one in which citations can be misquoted or obscured. Some "structural" papers do not contain diffraction data.

Iron compounds are important from both a chemical and biochemical point of view, and structural information is important for the understanding of the mechanistic role of iron ion chemical and biological reactions. It is hoped that this review will serve to crystallize the majority of relevant structural information of iron in a systematically useful manner and to highlihgt the similarities and differences which may be important and serve to stimulate further studies in this rich area of chemistry.

A related review of the structural chemistry of iron organometallic compounds is currently in progress.

10. ACKNOWLEDGEMENTS

The autors wish to thank those who gave plamission for reproduction of original figures and the Ministry of Education of the Slovak Republik for financial support (grant number 1/6106/99.

11.REFERENCES

1. M. Melník, I. Ondrejkovièová, V. Vanèová and C.E. Holloway, *Rev. Inorg. Chem.*, **17**, 55 (1997).
2. M. Melník, I. Ondrejkovièová, V. Vanèová and C.E. Holloway, *Rev. Inorg. Chem.*, **18**, 1 (1998).
3. H.B. Chin and R. Bau, *J. Am. Chem. Soc.*, **98**, 2434 (1976).
4. R.G. Teller, R.G. Finke, J.P. Collman, H.B. Chin and R. Bau, *J. Am. Chem. Soc.*, **99**, 1104 (1977).
5. L.M. Clarkson, W. Clegg, D.C.R. Hockless and N.C. Novman, *Acta Cryst., Sect. C*, **48**, 236 (1992).
6. K.H. Pannel, Yu-sen Chen, K. Belknap, C.C. Wu, J. Bernal, M.W. Creswick and Hsu Nan Huang, *Inorg. Chem.*, **22**, 418 (1983).

7. V.G. Albano, A. Araneo, P.L. Bellon, G. Ciani and M. Manassero, *J. Organomet. Chem.*, **67**, 413 (1974).
8. J. Donohue and A. Caron, *Acta Cryst.*, **17**, 663 (1964).
9. R. Boese and D. Bläser, *Z. Kristallogr.*, **193**, 289 (1990).
10. D. Braga, F. Grepioni and A.G. Orpen, *Organometallics*, **12**, 1481 (1993).
11. S.A. Goldfield and K.N. Raymond, *Inorg. Chem.*, **13**, 710 (1974).
12. R. Boese, D. Bläser and W. Petz, *Z. Naturforsch.*, **43b**, 945 (1988).
13. G.W. Harris, J.C.A. Boeyens and N.J. Coville, *Acta Cryst., Sect. C,* **39**, 1180 (1983).
14. M.B. Smith and R. Bau, *J. Am. Chem. Soc.*, **95**, 2388 (1973).
15. E. Bill, N. Blaes, K.F. Fischer. U. Gonser, K.H. Pauly, R. Preston, F. Seel, R. Staab and A.X. Trautwein, *Z. Naturforsch.*, **39b**, 333 (1984).
16. F.A. Cotton and J.M. Troup, *J. Am. Chem. Soc.*, **96**, 3438 (1974).
17. A.N. Nesmeyanov, L.V. Rybkin, N.A. Stelzer, Yu.T. Struchkov, A.S. Botsanov and M.I. Rybinskaya, *J. Organomet. Chem.*, **182**, 399 (1979).
18. C. Krüger, *Chem. Ber.*, **106**, 3230 (1973).
19. E. Niecke, H.J. Meternich, M. Nieger, D. Gudat, P. Wenderoth, W. Malisch, Ch. Halmer and W. Rucht, *Chem. Ber.*, **126**, 1299 (1993).
20. V.L. Goedken, Shie-Ming Peng, Jo Ann Molin-Norris and Yong-ae Park, *J. Am. Chem. Soc.*, **98**, 8391 (1976).
21. M. Ray, A.P. Golombek, M.P. Hendrich, V.G. Young,Jr. and A.S. Borovik, *J. Am. Chem. Soc.*, **118**, 6084 (1996).
22. J. Pickardt, L. Rösch and H. Schuman, *J. Organomet. Chem.*, **107**, 241 (1976).
23. B.T. Kilbourn, N.A. Raeburn and D.D. Thompson, *J. Chem. Soc. A,* 1906 (1969).
24. P.E. Riley and R.E. Davis, *Inorg. Chem.*, **19**, 159 (1980).
25. J.A.S. Howell , M.G. Palin, P. McArdle, Z. Goldschmidt, H.E. Gotlieb and D. Heroni-Langerman, *Inorg. Chem.*, **30**, 4683 (1991).
26. J.A.S. Howell , M.G. Palin, P. McArdle, D. Cunningham, Z. Goldschmidt, H.E. Gotlieb and D. Heroni-Langerman, *Inorg. Chem.*, **32**, 3493 (1993).
27 A.M.G. Dias Rodrigues, J.R. Lechat and R.H.P. Francisko, *Acta Cryst., Sect. C,* **48**, 159 (1992).
28. R.L. Keiter, A.L. Reingold, J.L. Hamerski and Ch. K. Castle, *Organometallics,* **3**, 1044 (1984).
29. G.L. Wood, E.N. Duesler, R.T. Paine and H. Noth, *Phosphorus, Sulphur and Silica,* **41**, 267 (1989).
30. A.H. Cowley, J.E. Kilduff, J.G. Lasch, N.C. Norman, M. Pakulski, F. Ando and T.C. Wright, *J. Am. Chem. Soc.*, **105**, 7751 (1983); *Organometalics,* **3**, 1044 (1984).
31. F.W.B. Einstein and R.D.G. Jones, *J. Chem. Soc. , Dalton Trans.*, 442 (1972).
32. A.H. Cowley, R.L. Geerts and Ch.M. Nunn, *J. Am. Chem. Soc.*, **109**, 6523 (1987).
33. H.G. Ang, Z.M. Cai, L.L. Koh and W.L. Kwik, *J. Chem. Soc., Chem. Commun.*, 850 (1991).
34. W.S. Sheldrick, S. Norton and O.Stelzen, *Z. Anorg. Allg. Chem.*, **475**, 232 (1981).
35. D. Dou, B. Kaufmann, E.N. Duesler, Tuqiang Chen, R.T. Paine and H. Nöth, *Inorg. Chem.*, **32**, 3056 (1993).
36. A.R. Barron, A.H. Cowley and Ch. Nunn, *Acta Cryst., Sect. C,* **44**, 750 (1988).
37. A.H. Cowley, R.E. Davis and K. Remandca, *Inorg. Chem.*, **20**, 2146 (1981).

38. A.H. Cowley, R.A. Kemp, E.A.V. Ebsworth, D.W.H. Rankin and M.D. Walkinshaw, *J. Organomet. Chem.*, **265**, C19 (1984).
39. D.W. Bennett, R.J. Neustadt, R.V. Parry and F.W. Cagle, *Acta Cryst., Sect. B,* **34**, 3362 (1978).
40. R.H. Nelson, R.J. Thoma, I. Vickovic and W.H. Watson, *Organometallics,* **3**, 1132 (1984).
41. Xiaoyong Sun, E.H. Wong, M.M. Turnbull, A.L. Rheingold, B.E. Waltermire and R.L. Ostrander, *Organometallics,* **14**, 83 (1995).
42. H. Schumann, M. Schäfers, H. Hemling and R. Weimann, *Z. Anorg. Allg. Chem.*, **622**, 1487 (1996).
43. E.O. Fischer, J. Schneider and D. Neugebauer, *Angew. Chem. Int. Ed. Engl.,* **23**, 820 (1984).
44. Ch. P. Casey, D.R. Mc Alister, M.W. Meszaros and K.J. Haller, *Cryst. Struct. Commun.,* **11**, 1015 (1982).
45. Ch. P. Casey, M.W. Meszaros, S.M. Neumann, I. Gennick Cesa and K.J. Haller, *Organometallics,* **4**, 143 (1985).
46. P.B. Hitchcock, M.F. Lappert, S.A. Thomas, A.J. Thorne, A.J. Carty and N.J. Taylor, *J. Organomet. Chem.*, **315**, 27 (1986).
47. F.A. Cotton, J.R. Kolb and B.R. Stults, *Inorg. Chim. Acta,* **15**, 239 (1975).
48. P.R. Jansen, A. Oskam and K. Olie, *Cryst. Struct. Commun.,* **4**, 667 (1975).
49. B.P.S. Chauhan, R.J.P. Corrin, G.F. Lanneau, Ch. Priou, N. Auner, H. Handwerker and E. Herdtweck, *Organometallics,* **14**, 1657 (1995).
50. N. Auner, C. Wagner, E. Herdtweck, M. Heckel and W. Hiller, *Bull. Soc. Chim. Fr.,* **132**, 599 (1995).
51. Ch. Zybill and G. Müller, *Organometallics,* **7**, 1368 (1988).
52. Ch. Zybill, D.L. Wilkinson, Ch. Leis and G. Müller, *Angew. Chem.,* **101**, 206 (1989).
53. Myong Euy Lee, Joan Soo Han and Chang Hwan Kim, *Bull. Korean Chem. Soc.,* **15**, 335 (1994).
54. Ch. Leis, Ch. Zybill, J. Lackmann and G. Müller, *Polyhedron,* **10**, 1163 (1991).
55. P.B. Hitchcock, J.A. Johnson and J.F. Nixon, *Angew. Chem. Int. Ed. Engl.,* **32**, 103 (1993).
56. R.F. Bryan and W.C. Schmidt, *J. Chem. Soc., Dalton Trans.,* 2337 (1974).
57. J.M. Cassidy and K.H. Whitmire, *Inorg. Chem.,* **30**, 2788 (1991).
58. P.P.M. de Lange, M.J.A. Kraakman, M. van Wijnkoop, H.F. Frühazf, K. Vrieze, W.J.J. Smeets and A.L. Spek, *Inorg. Chim. Acta,* **196**, 151 (1992).
59. R. Glaser, P.E. Haney and Ch.L. Barnes, *Inorg. Chem.,* **35**, 1758 (1996).
60. D. Ginderow, *Acta Cryst., Sect. B,* **30**, 2798 (1974).
61. R. Glaser, Young-Hee Yoo, G. Shiahuy Chen and Ch.L. Barnes, *Organometallics,* **13**, 2578 (1994).
62. H.P. Lane, S.M. Godfrey, Ch.A. Mc Auliffe and R.G. Pritchard, *J. Chem. Soc., Dalton Trans.,* 3249 (1994).
63. J.H. Mac Neil, A.C. Chiverton, S. Fortier, M.C. Baird, R.C. Hynes, A.J. Wiliams, K.F. Preston and T. Ziegler, *J. Am. Chem. Soc.,* **113**, 9834 (1991).
64. A.H. Cowley, R.E. Davis, M. Lattman, M. McKee and K. Remadna, *J. Am. Chem. Soc.,* **101**, 5090 (1979).

65. D.W. Bennet, D.S. Grubinska, S.E. Cremer and A.C. Peterson, *J. Cryst. Spectr. Res.,* **22**, 83 (1992).
66. D.A. Allison, J. Clardy and J.G. Verkade, *Inorg. Chem.,* **11**, 2804 (1972).
67. Lung-Shiang Luh and Ling Kang Liu, *Inorg. Chim. Acta,* **206**, 89 (1993).
68. J.A. Clucas, R.H. Dawson, P.A. Dolby, M.M. Harding, K. Pearson and A.K. Smith, *J. Organomet. Chem.,* **311**, 153 (1986).
69. L.P. Battaglia, D. Selledone, M. Nardeli, C. Pelizzi, G. Predieri and G.P. Chiusoli, *J. Organomet. Chem.,* **330**, 101 (1987).
70. D. Fenske, *Chem. Ber.,* **112**, 363 (1979).
71a. F.W.B. Einstein and Chung-Hsi Huang, *Acta Cryst., Sect. B,* **34**, 1486 (1978).
71b. Ling-Kang Liu, Lung-Shiang Luh and Han-Mou Gau, *Inorg. Chem.,* **31**, 3434 (1991).
72. E.C. Horsfield, D.W. Engel and K.G. Moodley, *Acta Cryst., Sect. C,* **44**, 80 (1988).
73. Chunbang Li, E.D. Stevens and S.P. Nolan, *Organometallics,* **14**, 3791 (1995).
74. J. Hein, M. Nieger and F. Niecke, *Organometallics,* **10**, 13 (1991).
75. D. Touchard, J.L. Fillaut, P.H. Dixneuf and L. Toupet, *J. Organomet. Chem.,* **317**, 291 (1986).
76. P. Conway, A.R. Manning and F.S. Stephens, *J. Organomet. Chem.,* **186**, C64 (1980).
77. D.S. Brown and G.W. Bushnel, *Acta Cryst.,* **22**, 296 (1967).
78. A.B. Burg., *Inorg. Chem.,* **25**, 4751 (1986).
79. H. Berke, W. Bankhart, G. Hutner, J. Von Seyerl and L. Zsolnai, *Chem. Ber.,* **114**, 2754 (1981).
80. L.P. Battaglia, G.P. Chiusoli, M. Nardeli, C. Pelizzi and G. Predieri, *Gaz. Chim. Ital.,* **116**, 207 (1986).
81. J.J. Brunet, R. Chaurin, G. Commeneges, B. Donnadieu and P. Leglaye, *Organometallics,* **15**, 1752 (1996).
82. C.E. Ash, Ch.M. Kim, M.Y. Darensbourg and A.L. Rheingold, *Inorg. Chem.,* **26**, 1357 (1987).
84. J.W. Connolly, A.H. Cowley and Ch.M. Nunn, *Polyhedron,* **9**, 1337 (1990).
85. U. Bodensieck, P. Braunstein, W. Deck, T. Faure, M. Knon and Ch. Stern, *Angew. Chem. Int. Ed. Engl.,* **33**, 2440 (1994).
86. Dao-Hinh Nguyen, Hua-Fen Hsu, M. Millar and S.A. Koch, *J. Am. Chem. Soc.,* **118**, 8963 (1996).
87. J.L. Roustan, N. Ansari, J.P. Charland and Y. Le Page, *Can. J. Chem.,* **67**, 2016 (1989).
88. D. Touchard, J.L. Fillaut, P.H. Dixneuf, C. Meali, M. Sabat and L. Toupet, *Organometallics,* **4**, 1684 (1985); D. Touchard, J.L. Fillaut, D.V. Khasnis, P.H. Dixneuf, C. Meali, D. Masi and L. Toupet, *ibid,* **7**, 971 (1988).
89a. W. Petz and F. Weller, *Z. Naturforsch.,* **46b**, 971 (1991).
89b. D. Touchard, J.L. Fillaut, D.V. Khasnis, P.H. Dixneuf, C. Mealli, D. Masi and L. Toupet, *Organometallics,* **7**, 67 (1988).
90. F.R. Ahmed, J.L.A. Roustan and M.Y. Al-Janabi, *Inorg. Chem.,* **24**, 2526 (1985).
91. R. Birk, H. Berke, H.U. Hund, G. Huttner, L. Zsolnai, L. Dahlenburg, U. Behrens and T. Sieliseh, *J. Organomet. Chem.,* **372**, 397 (1989).

92. B.L. Haymore and J.A. Ibers, *Inorg. Chem.*, **14**, 1369 (1975).
93a.P Conway, S.M. Grant, A.R. Manning and F.S. Stephens, *Inorg. Chem.*, **22**, 3714 (1983); *J. Organomet. Chem.*, **186**, C61 (1980).
93b.H. Le Bozec, P.H. Dixnuf, A.J. Carty and N.J. Taylor, *Inorg. Chem.*, **17**, 2568 (1978).
94. I.P. Lorenz, W. Hiller and M. Conrad, *Z. Naturforschung.*, **40b**, 1383 (1985).
95. A.J. Carty, P.H. Dixneuf, A. Gorgues, F. Hardstock, H. Le Bozee and N.J. Taylor, *Inorg. Chem.*, **20**, 3929 (1981).
96. M. Cygler, F.R. Ahmed, A. Forgues and J.L.A. Rouston, *Inorg. Chem.*, **22**, 1026 (1983).
97. M. Jänicke, H.U. Hund and H. Berke, *Chem. Ber.*,**124**, 719 (1991).
98. M. Knorr, S. Gilbert and M. Schubert, *J. Organomet. Chem.*, **347**, C17 (1988).
99. M. Melnik, *Coord. Chem. Rev.*, **47**, 239 (1982).
100. Shie-Ming Peng and J.A. Ibers, *J. Am. Chem. Soc.*, **98**, 8032 (1976).
101a. C. Slebodnik, J.C. Fettinger, H.B. Peterson and J.A. Ibers, *J. Am. Chem. Soc.*, **118**, 3216 (1996).
101b.K. Kim and J.A. Ibers, *J. Am. Chem. Soc.*, **113**, 6077 (1991).
102. L. Ricard, R. Weiss and M. Momenteau, *J. Chem. Soc., Chem. Commun.*, 818 (1986).
103. C. Tetreau, D. Lavalette, M. Momenteau, J. Fischer and R. Weiss, *J. Am. Chem. Soc.*, **116**, 11840 (1994).
104. L.E. McCandlish, B.D. Santarsiero, N.J. Rose and E.C. Lingafelter, *Acta Cryst., Sect. C*, **35**, 3053 (1979).
105. R.G. Litle, J.A. Ibers and J.E. Baldwin, *J. Am. Chem. Soc.*, **97**, 7049 (1975).
106. C. Slebodnik, M.L. Duval and J.A. Ibers, *Inorg. Chem.*, **35**, 3607 (1996).
107. D.G.A. Harshani de Silva, D.B. Leznof, G. Impey. I. Vernik, Zhe Jin and D.V. Stines, *Inorg. Chem.*, **34**, 4015 (1995).
108. D. Sellmann, G. Mahr, F. Knoch and M. Moll, *Inorg. Chim. Acta*, **224**, 45 (1994).
109. D. Sellmann, G. Binker, M. Moll and E. Hedtwek, *J. Organomet. Chem.*, **327**, 403 (1987).
110. D.K. Sivastava, N.P. Rath, L. Barton J.D. Ragini, O. Hollander, R. Godfroid and S.G. Shore, *Organometallics*, **12**, 2017 (1993).
111. M. Mangron, W.R. Clayton, O. Hollander and S.G. Shore, *Inorg. Chem.*, **16**, 2110 (1977).
112. Xiaoming He and J.F. Hartwig, *Organometallics*, **15**, 400 (1996).
113. A. Galindo, C. Mealli, J. Cuyas, D. Miguel, V. Riera, J.A. Perez- Martinez, C. Bois and Y. Jeannin, *Organometallics*, **15**, 2735 (1996).
114. V. Riera, J. Ruiz and A. Tiripicchio-Camellini, *J. Organomet. Chem.*, **327**, C5 (1987).
115. L. Vancea, M.J. Bennett, C.E. Jones, R.A. Smith and W.A.G. Graham, *Inorg. Chem.*, **16**, 897 (1977).
116. F.H. Carré and J.J.E. Moreau, *Inorg. Chem.*, **21**, 3099 (1982).
117. A.H. Cowley, N.C. Norman, M. Pakulski, D.L. Bricker and D.H. Russel, *J. Am. Chem. Soc.*, **107**, 8211 (1985).
118. J.R. Eveland, K.H. Whitmire and J.X. Saillard, *Inorg. Chem.*, **35**, 4400 (1996).

119. D. Sellmann, R. Weiss, F. Knoch, G. Ritter and J. Dengler, *Inorg. Chem.*, **29**, 4107 (1990).
120. D. Sellmann, G. Lanzrath, G. Huttner, L. Zsolnai, C. Krüger and K.H. Claus, *Z. Naturforsch.*, **38b**, 961 (1983).
121. D. Sellmann, H.E. Jonk, H.R. Pfeil, G. Huttner and J. von Seyerl, *J. Organomet. Chem.*, **191**, 171 (1980).
122a. D. Sellmann, G. Freiberger, R. Eberlein, E. Rőhlen, G. Huttner and L. Zsolnai, *J. Organomet. Chem.*, **323**, 21 (1987).
122b. D. Sellmann, T. Becker and F. Knoch, *Chem. Ber.*, **129**, 509 (1996).
123. F.A. Cotton, B.A. Frenz and A. Shaver, *Inorg. Chim. Acta,* **7**, 161 (1973).
124. Z. Nagy-Magos, L. Markó, A. Szakács-Schmidt, G. Gervasio, E. Belluroso and S.F. Kettle, *Bull. Soc. Chim. Belg.*, **100**, 445 (1991).
125. Wen-Feng Liaw, Ming-Hsi Chiang, Chen-Ju Liu, Piin-Jye Harn and Lin-Kang Liu, *Inorg. Chem.*, **32**, 1536 (1993).
126a. F. Galderazzo, G. Pampaloni, D. Vitali, G. Pellizzi, I. Collamati, S. Fredianni and M. Serra, *J. Organomet. Chem.*, **191**, 217 (1980).
126b. W.R. Scheidt, K.J. Haller, M. Fons, T. Mashiko and Ch.A. Reed, *Biochemistry,* **20**, 3653 (1981).
127. D. Sellmann, H. Kunstmann, F. Knoch and M. Moll, *Inorg. Chem.*, **27**, 4183 (1988).
128. D. Sellmann, T. Becker, T. Hofmann, F. Knoch and M. Moll, *Inorg. Chim. Acta,* **219**, 75 (1994).
129. G. Albertini, S. Antoniutti, M. Lanfranchi, G. Pelizzi and E. Bordignom, *Inorg. Chem.*, **25**, 950 (1986).
130. F.R. Askham, S.E. Saum and G.G. Stanley, *Organometallics,* **6**, 1370 (1987).
131. G. D. Fallon and b.M. Gatehouse, *Cryst. Struct. Commun.*, **11**, 1759 (1982).
132. H. Berke, G. Huttner and L. Zsolnai, *Chem. Ber.*, **114**, 3549 (1981).
133a. H. Berke, G. Huttner, G. Weiler and L. Zsolnai, *J. Organomet. Chem.*, **219**, 353 (1981).
133b. C.Y. Lee., Y. Wang and C.S. Liu, *Inorg. Chem.*, **30**, 3893 (1991).
134a. Wen-Feng Liaw, Ming-Hsi Chiang, Chia-Huei Lai, Show-Jen Chiou, Gene-Hsiang Lee and Shie-Ming Peng, *Inorg. Chem.*, **33**, 2493 (1994).
134b. Wen-Feng Liaw, Chia-Huei Lai, Ming-Iisi Chiang, Chen-Kang Hsieh, Gene- Hsiang Lee and Shie-Ming Peng, *J. Chinese Chem. Soc.*, **40**, 437 (1993).
135a. P. Braustein, S. Coco Cea, A. De Cian and J. Fischer, *Inorg. Chem.*, **31**, 4203 (1992).
135b. J. Takács, L. Markó, P. Kiprof, E. Herdtweck and W.A. Herman, *Polyhedron,* **8**, 1503 (1989).
136. A.E. Mauro, O.L. Casagrande, Jr., V.M. Nogulira, R.H.A. Santos, M.T.P. Gambardella, J.R. Lechat and M.F.J. Filho, *Polyhedron,* **12**, 297 (1993).
137. C.O. Kienitz, C. Thöne and P.G. Jones, *Inorg. Chem.*, **35**, 3990 (1996).
138. J. Takács, L. Markó and L. Párkányi, *J. Organomet. Chem.*, **361**, 109 (1989).
139. Wen-Feng Liaw, Der-Shiaw Ou, Yih-Chern Horng, Chia-Huei Lai, Gene-Hsiang Lee and Shie-Ming Peng, *Inorg. Chem.*, **33**, 2495 (1994).
140a. C.G. Brinkley, J.C. Dewan and M.S. Wrighton, *Inorg. Chim. Acta,* **121**, 119 (1986).
140b. J.W. Connolly, M.J. Hatler, A.H. Cowley and P.R. Sharp, *Polyhedron,* **10**, 841 (1991).

141. T.A. George, D.J. Rose, Yuanda Chang, Qin Chen and J. Zubieta, *Inorg. Chem.*, **34**, 1295 (1995).
142. S.A Wander, J.H. Reibenspies, Jang Sub Kim and M.Y. Darensbourg, *Inorg. Chem.*, **33**, 1421 (1994).
143. M.J. Begley, S.G. Puntambekar and A.H. Wright, *J. Organomet. Chem.*, **329**, C7 (1987).
144. R. Birk, H. Berke, G. Huttner and L. Zsolnai, *J. Organomet. Chem.*, **309**, C18 (1986).
145. T.M. Martin, P.T. Wood, G.L. Schimek, W.T. Pennington and J.W. Koles, *Inorg. Chem.*, **34**, 4385 (1995).
146. T.P. Fehlner, J. Ragaini, M. Mangion and S.G. Shore, *J. Am. Chem. Soc.*, **98**, 7085 (1976).
147. P. Paetzold, K. Delpy and R. Boese, *Z. Naturforsch.*, **43b**, 839 (1988).
148. B. Kaufmann, H. Nöth, R.T. Paine, K. Polborn and M. Thomann, *Angew. Chem. Int. Ed. Engl.*, **32**, 1446 (1993).
149. S.L. Lee, C.B. Knobler and H. Vahrenkamp, *Organometallics*, **10**, 670 (1991).
150. P.E. Baikie and O.S. Mils, *Inorg. Chim. Acta*, **1**, 55 (1967).
151. S.E. Kabin, M. Rauf and H. Varenkampf, *J. Organomet. Chem.*, **512**, 260 (1996).
152. H. Kisch, P. Reisser and F. Knoch, *Chem. Ber.*, **124**, 1143 (1991).
153. R.J. Doedens, *Inorg. Chem.*, **7**, 2323 (1968).
154. A.S. Kalugin, A.A. Pasynskii, I.L. Eremenko, E.A. Vasutinskaya, Yu. T. Struchkov and A.I. Yanovski, *J. Organomet. Chem.*, **386**, 225 (1990).
155. G. Le Borgne and D. Grandjean, *Acta Cryst., Sect.B*, **37**, 1198 (1991).
156. G. Gervasio, P.L. Stanghellini and P. Rossetti, *Acta Cryst., Sect. B*, **37**, 1198 (1981).
157. K.H. Jogun and J.J. Steyowski, *Acta Cryst.,Sect. B*, **35**, 2306 (1979).
158. R. Meij, D.J. Stufkens, K. Vrieze, A.M.F. Brouwers, J.D. Schagen, J.J. Zwinselman, A.R. Overbeek and C.H. Stam, *J. Organomet. Chem.*, **170**, 327 (1979).
159. B.G. Andrianov and Yu.T. Struchkov, *Koord. Khim.*, **4**, 937 (1978).
160. P. Bladon, M. Dekker, G.R. Knot, D. Willison, G.A. Jaffari, R.J. Doedens and K.W. Muir, *Organometallics*, **12**, 1725 (1993).
161. G.G. Aleksandrov, S.D.Alekseeva, N.S. Nametkin, A.I. Neklaev, B.I. Kobobkov, N.A. Parpiev, M.T. Tashev, V.D. Tjurin and H.B. Dustov, *Koord. Khim.*, **13**, 410 (1987).
162. A.I. Kolobkov, G.G. Aleksandrov, M.T. Tashev and H.B. Dustov, *Izv. Akad. Nauk SSSR, Ser. Khim.*, 698 (1987) ; A.I. Nehaev, B.I. Golobkov, G.G. Aleksandrov. M.T. Tashev, H.B. Dustov and N.A. Parpiev, *Koord. Khim.*, **14**, 416 (1988).
163. G.P. Khare and R.J. Doedens, *Inorg. Chem.*, **15**, 86 (1976).
164. D. Seyferth, G.B. Womack, Li-Cheng Song, M. Cowie and B.W. Hames, *Organometallics*, **2**, 928 (1983).
165. A. Shaver, P.J. Fitzpatrick, K. Steliou and I.S. Butler, *J. Am. Chem. Soc.*, **101**, 1313 (1979).
166. D. Seyferth, G.B. Womack, Li-Cheng Song, M. Cowie and B.W. Hames, *Organometallics*, **2**, 928 (1983).
167. M.R. Adams, J. Gallucci, A. Wojcicki and G.L. Long, *Inorg. Chem.*, **31**, 2 (1992).

168. N.S. Nametkin, B.I. Kolobkov, V.D. Tyurin, A.N. Muratov, A.I. Nekhaev, M. Mavlonov, A.Ya. Sideridu, G.G. Aleksandrov, A.V. Lebedev, M.T. Tashev and H.B. Dustov, *J. Organomet. Chem.*, **276**, 393 (1984).
169. A.I. Nekhaev, S.D. Alekseeva, N.S. Nametkin, V.D. Tyurin, B.I. Kolobkov, G.G. Aleksandrov, N.A. Parpiev, M.T. Tashev and H.B. Dustov, *J. Organomet. Chem.*, **297**, C33 (1985) ; G.G. Aleksandrov, A.I. Nekhaev, B.I. Golobkov, M.T. Tashev, H.B. Dustov, V.D. Tyurin and N.S. Nametkin, *Dokl. Akad, Nauk SSSR,* **289**, 880 (1986).
170. E.A. Chernyshev, O.V. Kuzmin, A.V. Lebedev, A.I. Gusev, M.G. Los, N.V. Alekseev, N.S. Nametkin, V.D. Tyurin, A.M. Krapivin, N.A. Kubasov and V.G. Zaikin, *J. Organomet. Chem.*, **252**, 143 (1983).
171. B. Hansert and H. Varenkamp, *J. Organomet. Chem.*, **459**, 265 (1993).
172. A.I. Nekhaev, S.D. Alekseeva, B.I. Kolobkov, G.G. Aleksandrov, M.T. Toshev, H.B. Dustov and N.A. Parpiev, *J. Organomet. Chem.*, **401**, 65 (1991).
173. D. Seferth, G.B. Womak, C.M. Archer and J.C. Dewan, *Organometallics,* **8**, 430 (1989).
174. A. Kramer, R. Lingnau, I.P. Lorenz and H.A. Mayer, *Chem. Ber.*, **123**, 1821 (1990).
175. D.L. Huges and D.R. Paulson, *Inorg. Chim. Acta,* **120**, 191 (1986).
176. A. Shaver, P.J. Fitzpatrick, K. Steliou and I.S. Butler, *J. Organomet. Chem.*, **172**, C59 (1979).
177. R.O. Gould and M.D. Walkinshaw, *Cryst. Struct. Commun.*, **10**, 1139 (1981).
178. H. Nöth and V. Rattay, *J. Organomet. Chem.*, **308**, 131 (1986).
179. W. Petz and F. Weller, *Organometallics,* **12**, 4056 (1993).
180. R.F. Bryan and H.P. Weber, *Acta Crystallogr. Sect. A,* **21**, 138 (1966).
181. P. Mathur, A.K. Dash, Md.M. Hossain, S.B. Umbarkar, C.V.V. Satyanarayana, Yu-Sheng Chen, E.M. Holt, S.N. Rao and M. Soriano, *Organometallics,* **15**, 1356 (1996).
182. J. Messelhäuser, I.P. Lorenz, K. Hang and W. Hiller, *Z. Naturforsch.*, **40b**, 1064 (1985).
183. G. Maier, D. Born, I. Bauer, R. Wolf, R. Boese and D. Cremer, *Chem. Ber.*, **127**, 173 (1994).
184. Xiao Lan Liu, Hong Li Chen, Fang Ming Miao, Li Cheng Song and Qing Mei Hu, *Acta Cryst., Sect. C,* **47**, 2199 (1991).
185. M.T. Ashby, *Inorg. Chem.*, **34**, 5429 (1995).
186. E.A. Chernyshev, O.V. Kuzmin, A.V. Lebedev, T.A. Mrachkovskaya, A.I. Gusev, I.I. Kirillova and A.V. Kisin, *Zh. Obsch. Khim.*, **56**, 367 (1986).
187. P. Mathur, Md. Munkhir Hossain and A.L. Reingold, *J. Organomet. Chem.*, **507**, 187 (1996).
188. M.D. Westmayer, T.B. Rauchfuss and A.K. Verma, *Inorg. Chem.*, **35**, 7140 (1996).
189. P.M. Treichel, W.K. Dean and J.C. Calabrese, *Inorg. Chem.*, **12**, 2908 (1973).
190. Yao Xin-Kan, Wang Ru-Ji, Wang Hong-Gen, Song Li-Cheng, Hu Qing-Mei and Wang Ji-Tao, *Acta Cryst., Sect C,* **45**, 575 (1989).
191. P. Mathur and Md. Munkir Hossain, *Organometallics,* **12**, 2398 (1993).
192. W. Henslee and R.E. Davis, *Cryst. Struct. Commun.*, **1**, 403 (1972).
193. H.B. Chin and R. Bau, *Inorg. Chem.*, **17**, 2314 (1978).

194. G. Dettlaf, P. Hübener, J. Klimes and E. Weiss, *J. Organomet. Chem.*, **229**, 63 (1982).
195. F.A. Cotton and J.M. Troup, *J. Chem. Soc., Dalton Trans.*, 800 (1974).
196. E. Delgado, E. Hernandez, O. Rossell, M. Seco, E.G. Puebla and C. Ruiz, *J. Organomet. Chem.*, **455**, 177 (1993).
197. P. Mathur, B. Manimaran, R. Trivedi, Md.M. Hossain and M. Arabatti, *J. Organomet. Chem.*,**515**, 155 (1996).
198. Xintao Wu, K.S. Bose, E. Sinn and B.A. Averill, *Organometallics,* **8**, 251 (1989).
199. M.A. Poraj-Koshits, A.S. Ancyshkina, S.B. Kacer, A.S. Ablullaev, A.S. Kalugin, I.L. Eremenko, A.S. Aliev and A.A. Pasynskij, *Zh. Vses, Khim., D.I. Mend.*, **32**, 102 (1979) ; S.B. Katcer, A.S. Ancyshkina, A.S. Kalugin and A.S. Abdulaev, *Koord. Khim.*, **13**, 1132 (1987).
200. P. Mathur, Md. Munkir Hossain, S. Umbarkar, C.V.V Satyanarayana and V.G. Puranik, *J. Organomet. Chem.*, **506**, 313 (1996).
201. A.J. Banister, I.B. Gorrel, W. Clegg and K.A. Jorgensen, *J. Chem. Soc., Dalton Trans.*, 2229 (1989).
202. W. Gaete, J. Ros, R. Yaòez, X. Solans, C. Miravitlles and M. Aguiló, *Inorg. Chim. Acta,* **119**, 55 (1986).
203. Wang Ruji, Song Licheng, Wang Honggan and Hu Qingnei, *Chem. J. Chin. Univ.*, **11**, 832 (1990).
204. A.J. Mayr, Hsun-Shan Lien, K.H. Pannell and L. Parkanyi, *Organometallics,* **4**, 1580 (1985).
205. P. Mathur, Md. Munkir Hossain, S. Umbarkar, C.V.V. Satyanarayana, S.S. Tavale and V.G. Puranik, *Organometallics,* **14**, 959 (1995).
206. P. Mathur, B. Manimaran, Md. Munkir Hossain, R. Shanbag, J. Murthy, I.S. Saranathan, C.V.V. Satyanarayana and M.M Bhadbhade, *J. Organomet. Chem.*, **490**, 173 (1995).
207. G. Gogarth and M.H. Lavender, *J. Chem. Soc., Dalton Trans.*, 143 (1993).
208. M. Nieger, G. Bruder and R. Appel, *Z. Kristallogr.*, **209**, 839 (1994).
209. Chin Hsuan Wei and L.F.Dahl, *Inorg. Chem.*, **4**, 1 (1965).
210. Rajib Lal De, *J. Organomet. Chem.*, **243**, 331 (1983).
211. N.J. Taylor, G.N. Mott and A.J. Carty, *Inorg. Chem.*, **19**, 560 (1980).
212. J. Schulze, R. Boese and G. Schmid, *Chem. Ber.*, **113**, 2348 (1980).
213. P. Mathur, R. Trivedi, Md.M. Hossain, S.S. Tavale and V.G. Purkanik, *J. Organomet. Chem.*, **491**, 291 (1995).
214. Li-Cheng Song, Chao-Guo Yan, Qing-Mei Hu, Ru-Ji Wang, T.C.W. Mak and Xiao-Ying Huang, *Organometallics,* **15**, 1535 (1996).
215. A. Winter, L. Zsolnai and G. Hunter, *J. Organomet. Chem.*, **250**, 409 (1983).
216. F.A. Cotton, B.E. Hanson, J.D. Jamerson and B.R. Stults, *J. Am. Chem. Soc.*, **99**, 3293 (1977).
217. F. Bitterer, D.J. Brauer, F. Dörrenbach, F. Gal, P.C. Knüppel, O. Stelzer, C. Krüger and Y.H. Tsay, *Z. Naturforsch.*, **46b**, 1131 (1991).
218. M.M. Van Dyk, P.H. Van Rooyen and S. Lotz, *Inorg. Chim. Acta,* **163**, 167 (1989).
219. F. Bitterer, D.J. Brauer, F. Dörenbach and O. Stelzer, *J. Organomet. Chem.*, **399**, C4 (1990).

220. Ch.F. Campana, F.Yip-Kwai Lo and L.F. Dahl, *Inorg. Chem.*, **18**, 3060 (1979).
221. G. Hogarth, M.H. Lavender and K. Shuki, *Organometallics,* **14**, 2325 (1995).
222. D. Buchlotz, G. Hutner, L. Zsolnai and W. Imhof, *J. Organomet Chem.*, **377**, 25 (1989).
223. L. Weber, E. Lücke and R. Boese, *Chem. Ber.*, **122**, 809 (1989).
224. Minghuey Shieh, Shie-Ming Peng and Gene Hsiang Lee, *Inorg. Chem.*, **34**, 2251 (1995).
225. P. Wisian-Neilson, K.D. Onan and D. Seyferth, *Organometallics,* **7**, 917 (1988).
226. Y.W. Li, M.G. Newton and E.M. Holt, *Inorg. Chem.*, **32**, 5720 (1986).
227. R.B. King, F.J. Wu and E.M. Holt, *Inorg. Che.*, **25**, 1733 (1986).
228. P. Mathur, V.D. Reddy and R. Bohra, *J. Organomet. Chem.*, **401**, 339, (1991).
229. K. Fischer and H. Vahrenkamp, *Z. Anorg. Allg. Chem.*, **475**, 109 (1981).
230. D. Seyferth, C.M. Archer, D.P. Ruschke, M. Cowie and R.W. Hilts, *Organometallics,* **10**, 3363 (1991).
231. N. Bahr, E. Beckmann, K. Mathaurer, D. Hunkler, M. Keller, H. Prinzbach and H. Vahrenkamp, *Chem. Ber.*, **126**, 429 (1993).
232. D. Seyferth, G.B. Womack, C.M. Archer, J.P. Fackler,Jr. and D.O. Macler, *Organometallics,* **8**, 443 (1989).
233. R.B. King, N.K.Bhattacharyya and E.M. Holt, *J. Organomet. Chem.*, **394**, 305 (1990).
234. J.M. Savariault, J.J. Bonnet, R. Mathieu and J. Galy, *C.R. Acad. Sc. Paris, Ser. C,* **284**, 663 (1977).
235. B.K. Das and M.G.Kanatzidis, *Inorg. Chem.*, **34**, 6505 (1995).
236. J.E. Elis and Yu-Sen Chen, *Organometallics,* **8**, 1350 (1989).
237. Y.W. Li, M.G. Newton, N.K. Bhattacharyya and R.B. King, *Inorg. Chem.*, **31**, 2069 (1992).
238. R.B. King, F.J. Wu, N.D. Sadanani and E.M. Holt, *Inorg. Chem.*, **24**, 4449 (1985).
239. H. Ch. Böttcher, H. Hartung, A. Krug and B. Walther, *Polyhedron,* **13**, 2893 (1994).
240. K.H. van Dijk, D.J. Stufkens, A. Askam, M. Rotteweel and D. Heijdenrijk, *Organometallics,* **6**, 1665 (1987).
241. F.A. Cotton and J.M. Troup, *J. Am. Chem. Soc.*, **96**, 1233 (1974).
242. V. Kumar, M.G. Newton and R.B. King, *J. Organomet. Chem.*, **502**, 61 (1995).
243. V. Kumar, D.W. Lee, M.G. Newton and R.B. King, *J. Organomet. Chem.*, **502**, 61 (1995).
244. A.M. Arif, A.H. Cowley and M. Pakulski, *J. Am. Chem. Soc.*, **107**, 2553 (1985).
245. G. Le Borgne and R. Mathieu, *J. Organomet. Chem.*, **208**, 201 (1981).
246. B.W. Eichhorn, R.C. Haushalter and J.S. Merola, *Inorg. Chem.*, **29**, 728 (1990).
247. R.T. Baker, J.C. Calabrese, P.J. Krusic, M.J. Therien and W,C. Trogler, *J. Am. Chem. Soc.,* **110**, 8392 (1988).
248. Y.W. Li, M.G. Newton and R.B. King, *J. Organomet. Chem.*, **488**, 63 (1995).
249. A. Albini, H. Kirsch, C. Krüger and An-Pei Chiang, *Z. Naturforsch.*, **37b**, 463 (1982).
250. Minghuey Shieh and Mei-Huey Shieh, *Organometallics,* **13**, 920 (1994).

251. D. Seyferth, T.G. Wood, J.P. Fackler, Jr. and A.M. Mazany, *Organometallics,* **3**, 1121 (1984).
252. B. Walther, H. Hartung, S. Bombirra, A. Krug and H.C. Böttcher, *Organometallics,* **13**, 172 (1994).
253. A.L. Rhiengold and M.E. Fountain, *Organometallics,* **3**, 1417 (1984).
254. R.E. Bachman and K.H. Whitmire, *Organometallics,* **12**, 1988(1993).
255. M.A. El-Hinnauri, E.J. Wucherer, R. Lal De and H. Vahrenkamp, *J. Organomet. Chem.,* **329**, 381 (1987).
256. G.M. Boron, J.E. Finholt, R.B. King, J.W. Bibber and J.H. Kim, *Inorg. Chem.,* **21**, 3790 (1982).
257. M.G. Newton, R.B. King, M. Chang and J. Gimeno, *J. Am. Chem. Soc.,* **99**, 2802 (1977).
258. W. Clegg and S. Morton, *Inorg. Chem.,* **18**, 1189 (1979).
259. F.A. Cotton, R.J. Haines, B.E. Hanson and J.C. Sekutowski, *Inorg. Chem.***17**, 2010 (1978).
260. J.T. Lin, Yung Chun Chou, Yeh E. Shih, Fong E. Hong, Yuh S. Wen, Su-Ching Lin and Mei-Man Chen, *J. Chem. Soc., Dalton Trans.,* 1791 (1995).
261. P. Frankhauser, F. Kuhlmann, A. Krämer, H. Pritzkov and W. Siebert, *Chem. Ber.,* **126**, 1291 (1993).
262. R.S. Simons and C.A. Tessier, *Acta Cryst., Sect. C,* **51**, 1997 (1995).
263. A.J. Carty, G.N. Mott and N.J. Taylor, *J. Organomet. Chem.,* **182**, C69 (1979).
264. F.A. Cotton and J.M. Troup, *J. Am. Chem. Soc.,* **96**, 4422 (1974).
265. P.B. Hitchcock, T.J. Madden and J.F. Nixon, *J. Organomet. Chem.,* **463**, 155 (1993).
266. Ch. Santelli-Rouvier, Ch. Coin, L. Toupet and M. Santelli, *J. Organomet. Chem.,* **495**, 91 (1995).
267. J. Meunier-Piret, P. Piret and M. Van Meerssche, *Bull. Soc. Chim. Belg.,* **76**, 374 (1967).
268. Wai Kwok Wong, Kwok W. Chiu, G. Wilkinson, M. Motevalli and M.B. Hursthouse, *Polyhedron,* **4**, 1231 (1985).
269. H. Ch. Böttcher, A. Krug and H. Hantung, *Polyhedron,* 14, 901 (1995).
270. H. Tobita, I. Shinagawa, S. Ohnuki, M. Abe, H. Izumi and H. Ogino, *J. Organomet. Chem.,* **473**, 187 (1994).
271. M.G. Newton, R.B. King, M. Chang and J. Gimeno, *J. Am. Chem. Soc.,* **100**, 326 (1978).
272. G. De Leeano, J.S. Field, R.J. Haines, B. Mc Culloch, L. Meintjes, Ch. Monberg, K.G. Moodley, G.M. Olivier, C.N. Sampson and N.D. Steen, *J. Organomet. Chem.,* **228**, C66 (1982).
273. Yuan-Fu Ju, A. Vojcicki, M. Calligaris and G. Nardin, *Organometallics,* **5**, 47 (1986).
274. M. Herbenhold, K. Schamel, G. Herrmann, A. Gieren, C. Ruiz-Perez and T. Hübner, *Z. Anorg. Allg. Chem.,* **562**, 49 (1988).
275. H. Vahrenkamp and G. Walters, *Angew. Chem.,* **95**, 152 (1983).
276. K.G. Moodley, D.W. Engel, J.S. Field and R.J. Haines, *Polyhedron,* **12**, 533 (1993).
277. R.B. King, N.K. Bhattacharyya and E.M. Holt, *Inorg. Chem.,* **30**, 1174 (1991).
278. R.S. Simons and C.A. Tessier, *J. Organomet. Chem.,* **15**, 2604 (1996).

279. T.M. Bockman, H.C. Cho and J.K. Kochi, *Organometallics,* **14**, 5221 (1995).
280. H.B. Chin, M.B. Smith, R.D. Wilson and R. Bau, *J. Am. Chem. Soc.,* **96**, 5285 (1974).
281. J.S. Field, R.J. Haines, E. Minshall, C,N. Sampson, J. Sundermayer, Ch.C. Allen and J.C.A. Boeyens, *J. Organomet. Chem.,* **309**, C21 (1986).
282. N.K. Bhattacharyya, T.G. Coffy, W. Quintana, T.A. Salupo, J.C. Bicker, T.B. Shay, M. Payne and S.G. Shore, *J. Organomet. Chem.,* **9**, 2368 (1990).
283. A.H. Cowley, N.C. Norman and M. Pakulski, *J. Am. Chem. Soc.,* **106**, 6844 (1984).
284. W. Clegg, *Inorg. Chem.,* **15**, 1609 (1976).
285. A.J. Schutz and R. Eisenberg, *Inorg. Chem.,* **12**, 518 (1973).
286. N.J. Taylor, M.S. Arabi and R. Mathieu, *Inorg. Chem.,* **19**, 1740 (1980).
287. F.J. Berry, Ch.J. Jones, J.A. Mc Cleverty, J. Sharpe, N.A. Bailey and M. Deadman, *J. Organomet. Chem.,* **353**, 209 (1988).
288. D. Sellmann, R. Weiss and F. Knoch, *Angew. Chem.,* **101**, 1719 (1989).
289. G.R. Davies, R.H.B. Mais, P.G. Owston and D.T. Thompson, *J. Chem. Soc. A,* 1251 (1968).
290. Jin Zhao, W.T. Pennington and J.W. Kolis, *J. Chem. Soc., Chem. Commun.,* 265 (1992).
291. R.E. Ginsberg, R.K. Rothrock, R.G. Tinke, J.P. Collman and L.F. Dahl, *J. Am. Chem. Soc.,* **101**, 6550 (1979).
292. T.M. Martin, G.L. Schimek, D.A. Mlsna and J.W. Kalis, *Phosphorus, Sulphur and Silicon,* **93-94**, 94 (1994).
293. A.M. Caminade, J.P.Majoral, M. Sanchez, R. Mathieu, S. Attali and A. Grand, *Organometallics,* **6**, 1459 (1987).
294. W. Deck and H. Vahrenkamp, *Z. Anorg. Allg. Chem.,* **598/599**, 83 (1991).
295. J.R. Eveland and K.H. Whitmire, *Angew. Chem., Int. Ed. Engl.,* **35**, 741 (1996).
296. R.A. Bartlett, H.V. Rasika Dias, K.M. Flynn, M.M. Olmstead and Ph. P. Power, *J. Am. Chem. Soc.,* **109**, 5699 (1987).
297. D.A. Du Bols, E.N. Duesier and R.T. Palne, *Organometallics,* **3**, 1913 (1984).
298. K.M. Flynn, R.A. Bartlett, M.M. Olmstead and Ph. P. Power, *Organometallics,* **5**, 813 (1986).
299. K.M. Flynn, H. Hope, B.D. Murray, M.M. Olmstead and Ph.P. Power, *J. Am. Chem. Soc.,* **105**, 7750 (1983).
300. G. Hogarth, *J. Organomet. Chem.,* **406**, 391 (1991).
301. K.M. Flynn, M.M. Olmstead and Ph. P. Power, *J. Am. Chem. Soc.,* **105**, 2085 (1983).
302. K.M. Flynn, B.D. Murray, M.M. Olmstead and Ph.P. Power, *J. Am. Chem. Soc.,* **105**, 7460 (1983).
303. D. Jacoby, C. Floriani, A. Chiesi-Villa and C. Rizzoli, *J. Chem. Soc., Dalton Trans.,* 813 (1983).
304. A. Ceriotti, L. Resconi, F. Demartin, G. Longoni, M. Manassero and M. Sansoni, *J. Organomet. Chem.,* **249**, C35 (1983).
305. P.L. Bogdan, M. Sabat, S.A. Sunshine, C. Woodcock and D.F. Shriver, *Inorg. Chem.,* **27**, 1904 (1988).
306. F. Ragaini, Jeong-Sup Song, D.L. Ramage, G.L. Geoffray, G.A.P. Yap and A.L. Rheingold, *Organometallics,* **14**, 387 (1995).

307a. Liu Oiwang, Hu Xiang, Liu Shutang, Su Haiquam, Wang Boyi and Wang Ming, *Wuji Huaxue Xuebao,(J. Inorg. Chem.),* **7**, 202 (1991).
307b. A. Gourdon and Y. Jeannin, *Organometallics,* **5**, 2406 (1986).
308. M.A. Andrews, G. van Buskirk, C.B. Knobler and H.D. Kaesz, *J. Am. Chem. Soc.,* **101**, 7245 (1979).
309. L.E. Crascall, B.H.S. Thimmappa, A.L. Rheingold, R. Ostander and T.P. Tehlner, *Organometallics,* **13**, 2153 (1994).
310. H. Adams, N.A. Bailey, G.W. Bentley and B.E. Mann, *J. Chem. Soc., Dalton Trans.,* 1831 (1989).
311. M.A. Andrews, C.B. Knobler and H.D. Kaesz, *J. Am. Chem. Soc.,* **101**, 7260 (1979).
312. P. Suter and H. Vahrenkamp, *Chem. Ber.,* **128**, 793 (1995).
313. W. Deck, A.K. Pawell and H. Vahrenkamp, *J. Organomet. Chem.,* **411**, 431 (1991).
314. H. Kisch, C. Krüger and A. Trautwein, *Z. Naturforsch.,* **36b**, 205 (1981).
315. Z. Zhigui, W. Lixin, Z. Hengbin, Y. Ling and F. Yuguo, *Eur. J. Solid State Inorg. Chem.,* **28**, 1269 (1991)
316. F. Yip-Kwai Lo, G. Longoni, P. Chini, L.D. Lower and L.F. Dahl, *J. Am. Chem. Soc.,* **102**, 7691 (1980).
317. G. Hogath, N.J. Taylor, A.J. Carty and A. Mayer, *J. Chem. Soc., Chem. Commun.,* 834 (1988).
318. F. Ragaini, D.L. Ramage, Jeong-Sup Song, G.L. Geoffroy and A.L. Rheingold, *J. Am. Chem. Soc.,* **115**, 12183 (1993).
319. R.E. Bachmann and K.H. Whitmire, *Inorg. Chem.,* **33**, 2527 (1994).
320. G.B. Karet, C.L. Stern, D.M. Norton and D.F. Shiever, *J. Am. Chem. Soc.,* **115**, 9979 (1993).
321. A. Bino, F.A. Cotton, P. Lahuerta, P. Puebla and R. Uson, *Inorg. Chem.,* **19**, 2357 (1980).
322. R.L. Holliday, L.C. Roof, B. Hargus, D.M. Smith, P.T. Wood, W.T. Pennington and J.W. Kolis, *Inorg. Chem.,* **34**, 4392 (1995).
323. J.C. Vites, Ch. Eigrenbrot and T.P. Fehlner, *J. Am. Chem. Soc.,* **106**, 4633 (1984).
324. S.N. Konchenko, A.V. Virovles, V.A. Varnek, S.V. Tkachev, N.V. Podbrezskaya and V.A. Maksakov, *Zh. Strukt. Khim.,* **37**, 337 (1996); Engl. Ed., p. 295
325. L.C. Roof, D.M. Smith, G.V. Drake, W.T. Pennington and J.W. Kolis, *Inorg. Chem.,* **34**, 337 (1995).
326. D. Lentz and R. Marschal, *Organometallics,* **10**, 1487 (1991).
327. F.A. Cotton and J.M. Troup, *J. Am. Chem. Soc.,* **96**, 5070 (1974).
328. D. Braga, F. Grepioni, L.J. Farrugia and B.F.G. Johnson, *J. Chem. Soc., Dalton Trans.,* 2911 (1994).
329. D. Braga, L. Farrugia, F. Grepioni and B.F.G. Johnson, *J. Organomet. Chem.,* **464**, C39 (1994).
330. Chin hsuan Wei and L.F. Dahl, *J. Am. Chem. Soc.,* **91**, 1351 (1969).
331. I. Brüdgam, H. Hartl and D. Lentz, *Z. Naturforsch.,* **39b**, 721 (1984).
332. G. Raper and W.S. Mc Donald, *J. Chem. Soc. A,* 3430 (1971).
333. F.A. Cotton and J.M. Troup, *J. Am. Chem. Soc.,* **96**, 4155 (1974).
334. D. Buchholz, G. Huttner and W. Imhof, *J. Organomet. Chem.,* **388**, 307 (1990).

335. P. Baistrocchi, D. Cauzzi, M. Lanfranchi, G. Predieri, A. Tiripicchio and M. Tiripicchio Camellini, *Inorg. Chim. Acta,* **235**, 173 (1995).
336. A. Winter, L. Zsolnai and G. Huttner, *Chem. Ber.,* **115**, 1286 (1982).
337. H. Adams, A.G. Carr, B.E. Mann and R. Melling, *Polyhedron,* **14**, 2771 (1995).
338. D.J. Dahm and R.A. Jacobson, *J. Am. Chem. Soc.,* **90**, 5106 (1968).
339. K. Knoll, G. Huttner, L. Zsolnai, B. Jibril and M. Vasiucionek, *J. Organomet. Chem.,* **294**, 91 (1985).
340. E. Iiskola, T.A. Pakkanen, T.T. Pakkanen and T. Venäläinen, *Acta Chem. Scan. A,* **37**, 125 (1983).
341. J.B. Murray, B.K. Nicholson and A.J. Whitton, *J. Organomet. Chem.,* **385**, 91 (1990).
342. L.F. Dahl and J.F. Blount, *Inorg. Chem.,* **4**, 1373 (1965).
343. R.B. King, G.S. Chorghade, N.K. Bhattacharyya, E.M. Holt and G.J. Long, *J. Organomet. Chem.,* **411**, 419 (1991).
344. G.B. Karet, D.M. Norton, Ch.L. Stern and D.F. Shriver, *Inorg. Chem.,* **33**, 5750 (1994).
345. H. Adams, Xuetai Chen and B.E. Mann, *J. Chem. Soc., Dalton Trans.,* 2159 (1996).
346. R. Bau, B. Don, R. Greatrex, R.J. Haines, R.A. Love and R.D. Wilson, *Inorg. Chem.,* **14**, 3021 (1975).
347. G.K. Chen, G.H. Cheng and T.H. Hseu, *Organometallics,* **6**, 868 (1987).
348. B.K. Das and M.G. Kanatzidis, *J. Organomet. Chem.,* **513**, 1 (1996).
349. K. Knoll, G. Huttner, L. Zsolnai, O. Orama and M. Vasiucionek, *J. Organomet. Chem.,* **310**, 225 (1986).
350. D.L. Sunick, P.S. White and C.K. Schauer, *Organometallics,* **12**, 245 (1993).
351. E. Keller and H. Vahrenkamp, *Chem. Ber.,* **114**, 1124 (1981).
352. G. Huttner, J. Schneider, G. Mohr and J. von Seyerl, *J. Organomet. Chem.,* **191**, 161 (1980).
353. V.D. Patel, A.A. Cherkas, D. Nucciarone, N.J. Taylor and A.J. Carty, *Organometallics,* **4**, 1792 (1985).
354. Ch.J. Cardin, D.J. Cardin, N.B. Kelly, G.A. Lawless and M.B. Power, *J. Organomet. Chem.,* **341**, 447 (1988).
355. D.J. Brauer, S. Hietkamp, H. Sommer, O. Stelzer, G. Müller and C. Krüger, *J. Organomet. Chem.,* **288**, 35 (1985).
356. D.J. Brauer, S. Hietkamp, H. Sommer and O. Stelzer, *J. Organomet. Chem.,* **281**, 187 (1985).
357. K.H. Whitmize, C.B. Lagrone and A.L. Rheingold, *Inorg. Chem.,* **25**, 2472 (1986).
358. T.M. Bockman and J.K. Kochi, *J. Am. Chem. Soc.,* **109**, 7725 (1987).
359. W. Clegg G.M. Sheldrick, D. Stalke, S. Bahaduri and H.K. Khwaja, *Acta Cryst., Sect. C,* **40**, 2045 (1984).
360. R.J. Doedens, *Inorg. Chem.,* **8**, 570 (1969).
361. A. Winter, L. Zsolnai and G. Huttner, *Z. Naturforsch.,* **37b**, 1430, (1982).
362. Jeong-Sup Song, G.L.Geoffroy and A.L.Rheingold, *Inorg. Chem.,* **31**, 1505 (1992).
363. Song. Hua, Li Dazhen, Hou Enjian and Deng Xixian, *Acta Phys. Chim. Sinica,* **6**, 730 (1990).
364. R.D. Adams and J.E. Babin, *Inorg. Chem.,* **25**, 3418 (1986).

365. G. Gervasio, *J. Organomet. Chem.*, **445**, 147 (1993).
366. L. Markó, B. Markó-Monostory, Th. Madach and H. Vahrenkamp, *Angew. Chem.* **92**, 225 (1980).
367. J.P. Fackler, Jr., T.G. Wood and Ch.F. Campana, *Inorg. Chim. Acta,* **82**, 31 (1984).
368. B. Eber, G. Huttner, L. Zsolnai and W. Imhof, *J. Organomet. Chem.*, **402**, 221 (1991).
369. R. Seidel, B. Schnautz and G. Henkel, *Angew. Chem. Int., Ed. Engl.*, **35**, 1710 (1996).
370. A. Winter, L. Zsolnai and G. Huttner, *J. Organomet. Chem.*, **234**, 337 (1982).
371. E. Lindner, G.A. Weiss, W. Hiller and R. Fawzi, *J. Organomet. Chem.*, **255**, 245 (1983).
372a. A. Winter, L. Zsolnai and G. Huttner, *J. Organomet. Chem.*, **232**, 47 (1982).
372b. W. Imhof, B. Eber, G. Huttner and Ch. Emmerich, *J. Organomet. Chem.*, **447**, 21 (1993).
373. H.H. Ohstand and J.K. Kochi, *J. Am. Chem. Soc.*, **108**, 2897 (1986).
374. S.L.Cook, J. Evans, L.R. Gray and M. Webster, *J. Organomet. Chem.*, **236**, 367 (1982).
375. M.T. Bautista, M.R. Jordan, P.S. White and C.K. Schauer, *Inorg. Chem.*, **32**, 5429 (1993).
376. J.K. Kouba, E.L. Metterties, M.R. Thompson and V.W. Day, *Organometallics,* **2**, 1065 (1983).
377. A.L. Rheingold, R.L. Ostrander and P. Mathur, *Acta Cryst., Sect. C,* **49**, 1741 (1993).
378. E.P. Kybà, R.E. Davis, K.L. Hassett, J.S. Mc Kennis and B. Sheikh, *Phosphorus and Sulfur,* **18**, 275 (1983).
379. E.P. Kyba, K.L. Hassett, B. Sheikh, J.S. Mc Kennis, R.B. King and R.E. Davis, *Organometallics,* **4**, 994 (1985).
380. D.J. Brauer, S. Hietkamp, H. Sommer and O. Stelzer, *Z. Naturforsch.*, **40b**, 1677 (1985).
381. J.R. Eveland, J.Y. Saillard and K.H. Whitmire, *Inorg. Chem.*, **36**, 330 (1997).
382. J. Doherty, A.R. Manning and F.S. Stephens, *Inorg. Chem.*, **21**, 3332 (1982).
383. W. Deck, M. Schwarz and H. Vahrenkamp, *Chem. Ber.*, **120**, 1515 (1987).
384. D.A. Lesch and Th.B. Rauchfuss, *Organometallics,* **1**, 499 (1982).
385. D.J. Brauer, S. Hietkamp, H. Sommer, O. Stelzer, G. Müller, M.J. Romao and C. Krüger, *J. Organomet. Chem.*, **296**, 411 (1985).
386. R.B. King, F.J. Wu and E.M. Holt, *J. Organomet. Chem.*, **314**, C27 (1986).
387. U. Berger and J. Strähler, *Z. Anorg. Allg. Chem.*, **516**, 19 (1984).
388. M.A. Walters and J.C. Devan, *Inorg. Chem.*, **25**, 4889 (1986).
389. R.E. Bachman, S.K. Miller and K.H. Whitmire, *Inorg. Chem.*, **33**, 2075 (1994).
390. R.E. Bachman, S.K. Miller and K.H. Whitmire, *Organometallics,* **14**, 796 (1995).
391. P. Henderson, M. Rossignoli, R.C. Burns, M.L. Scudder and D.C. Kraig, *J. Chem. Soc. Dalton Trans.*, 1641 (1994).
392. M. Ferrer, O. Rossell, M. Seca and P. Braunstein, *J. Organomet. Chem.*, **364**, C5 (1989).
393. Shifang Luo and K.H. Whitmire, *Inorg. Chem.*, **28**, 1424 (1989).
394. J.M. Cassidy and K.H. Whitmire, *Inorg. Cem.*, **28**, 2494 (1989).

395. G.W. Drake, G.L. Schimek and J.W. Kolis, *Inorg. Chim. Acta,* **240**, 63 (1995).
396. H. Westermann, M. Nieger, E. Niecke, J.P. Majoral, A.M. Caminade and R. Mathieu, *Organometallics,* **8**, 244 (1989).
397. B. Hansert, A.K. Powell and H. Vahrenkamp, *Chem. Ber.,* **124**, 2697 (1991).
398. G. Gervasio, R. Rossetti and P. Stanghellini, *J. Chem. Res.(S),* 334 (1979).
399. R.J. Doedens and L.F. Dahl, *J. Am. Chem. Soc.,* **88**, 4847 (1966).
400. R.D. Adams, J.E. Babin, J. Estrada, Jin-Guu Wang, M.B. Hall and A.A. Low, *Polyhedron,* **8**, 1885 (1989).
401. G. van Buskirk, C.B. Knobler and H.D. Kaesz, *Organometallics,* **4**, 149 (1985).
402. H. Bantel, A.K. Powell and H. Vahrenkamp, *Chem. Ber.,* **123**, 677 (1990).
403. B. Eber, G. Huttner, D. Grünauer, W. Imhof and L. Zsolnai, *J. Organomet. Chem.,* **414**, 361 (1991).
404. G. Fachnetti, G. Fochi, T. Funaioli and P.F. Zanazzi, *J. Chem. Soc., Chem. Commun.,* 89 (1987); C. Mealli, D.M. Proserpio, G. Fachinnetti, T. Funaioli, G. Fochi and P.F. Zanazzi, *Inorg. Chem.,* **28**, 1122 (1989).
405. S.N. Datta, R.K. Kondru and P. Mathur, *J. Organomet. Chem.,* **470**, 169 (1994).
406. T. Jaeger, S. Aime and H. Vahrenkamp, *Organometallics,* **5**, 245 (1986).
407. S. Attali, F. Daham, R. Mathieu, A.M. Camonade and J.P. Majoral, *J. Am. Chem. Soc.,* **110**, 1990 (1988).
408. U. Flörke and J. Haupt, *Z. Kristallographie,* **194**, 140 (1991).
409. H. Bantel, B. Hansert, A.K. Powell, M. Tasi and H. Vahrenkamp, *Angew. Chem.,* **101**, 1084 (1989).
410. D.E. Fjare and W.L. Glodfelter, *J. Am. Chem. Soc.,* **103**, 1572 (1981); *Inorg. Chem.,* **20**, 3533 (1981).
411. P. Zanello, F. Laschi, A. Cinquantini, R.D. Pergola, L. Garlaschelli, M. Cucco, F. Demartin and T.R. Spalding, *Inorg. Chim. Acta,* **226**, 1 (1994).
412. M. Taxchikawa, J. Stein, E.L. Mueterties, R.G. Teller, M.A. Beno, E. Gebut and J.M. Wiliams, *J. Am. Chem. Soc.,* **102**, 6648 (1980).
413. D. Buchholz, G. Huttner, W. Imhof and O. Orama, *J. Organomet. Chem.,* **388**, 321 (1990).
414. K. Whitmire, D.S. Shriver and E.M. Holt, *J. Chem. Soc., Chem. Commun.,* 780 (1980).
415. T.P. Fehlner, C.E. Housecroft, W.R. Scheidt and K.S. Wong, *Organometallics,* **2**, 825 (1983).
416. D.J. Brauer, G. Hasselkuss, S. Hietkamp, H. Sommer and O. Stelzer, *Z. Naturforsch.,* **40b**, 961 (1985).
417. B.I. Kolobkov, N.S. Nametkin, V.D. Tyurin, A.I. Nekhaev, G.G. Aleksandrov, M.T. Tashiev and H.B. Dustov, *J. Organomet. Chem.,* **301**, 349 (1986).
418. Li-Cheng Song, Oing-Mei Hu, Lu-Yan Zhang, He Wang, Zhong-Yuan Zhore and Li Liu, *J. Organomet. Chem.,* **412**, C19 (1991).
419. Li-Cheng Song, M. Kadiata, Ji-Tao Wang, Ru-Ji Wang and Hong-Gen Wang, *J. Organomet. Chem.,* **340**, 239 (1988).
420. J.M. Coleman, A. Wojcicki, Ph.J. Pollick and L.F. Dahl, *Inorg. Chem.,* **6**, 1236 (1967).

421. Li-Cheng Song, Chao-Guo Yan, Qing-Mei Hu and Xiao-Ying Huang, *J. Organomet. Chem.*, **505**, 119 (1995).
422. Li-Cheng Song, Qing-Mei Hu, Chao-Guo Yan, Ru-Ji Wang and T.C.W. Mak, *Acta Cryst., Sect. C,* **52**, 1357 (1996).
423. P. Mathur, R. Trivedi and C.V.V. Satyanarayana, *Organometallics,* **15**, 1062 (1996)
424. G. Huttner, G. Mohr, B. Pritzlaff, J. von Seyerl and L. Zsolnai, *Chem. Ber.*, **115**, 2044 (1982).
425. S.G. Anema, G.C. Barris, K.M. Mackay and B,K, Nicholson, *J. Organomet. Chem.*, **350**, 207 (1988).
426. K.S. Bose, E. Sinn and B.A. Averill, *Organometallics,* **3**, 1126 (1984).
427. D. Seyferth, A.M. Kiwan and E. Sinn, *J. Organomet. Chem.*, **281**, 111 (1985).
428. Song-Ping Huang and M.G. Kanatzidis, *Inorg. Chem.*, **32**, 821 (1993).
429. Li-Chenf Song and Qing-Mei Hu, *J. Organomet. Chem.*, **414**, 219 (1991).
430. P.V. Broadhurst, B.F.G. Johnson, J. Lewis and P.R. Raithby, *J. Chem. Soc., Chem. Commun.*, 140 (1982).
431. Li-Cheng Song, M. Kadiata, Ji-Tao Wang, Ru-Ji Wang and Hong-Gen Wang, *J. Organomet. Chem.*, **391**, 387 (1990).
432. B.I. Kolobkov, A.I. Nechaev, G.G. Aleksandrov, M.T. Tashev, H.B. Dustov and A.Ya. Sibemby, *Izv. Akad. Nauk. SSSR, Ser. Khim.*, 953 (1987).
433. Chang-Soo Jun, Xiangsheng Meng, K.J. Haller and Th.P. Fehlner, *J. Am. Chem. Soc.,* **113**, 3603 (1991).
434. A. Gourdon and Y. Jeannin, *J. Organomet. Chem.*, **304**, C1 (1986).
435. M. Cané, M.C. Iapalucci, G. Longoni, F. Demartin and L. Grossi, *Mater. Chem. and Physics,* **29**, 395 (1991).
436. K.H. Whitmire, J. Scott Leight, Shifang Luo, Minghuey Shieh, M.D. Fabiano and A.L. Rheingold, *New. J. Chem.*, **12**, 397 (1988).
437. R. Lal De and H. Vahrenkamp, *Angew. Chem.*, **96**, 961 (1984); *Z. Naturforsch.,* **41b**, 273 (1986).
438. Shifang Luo and K.H. Whitmire, *J. Organomet. Chem.,* **376**, 297 (1989).
439. E. Röttinger and H. Vahrenkamp, *Angew. Chem.,* **90**, 294 (1978); *J. Organomet. Chem.,* **213**, 1 (1981).
440. P. Mathur, A.K. Dash, Md. Munkir Hossain, C.V.V. Satyanarayana and B. Verghese, *J. Organomet. Chem.*, **506**, 307 (1996).
441. P. Mathur, A.K. Dash, Md. Minkir Hossain and M.F. Mahon, *J. Organomet. Chem.*, **471**, 185 (1994).
442. R. Hourihane, T.R. Spalding, G. Ferguson, T. Deeney and P. Zanello, *J. Chem. Soc., Dalton Trans.*, 43 (1993).
443. A. Gourdon and Y. Jeannin, *C.R. Acad. Sc. Paris, Serie II,* **295**, 1101 (1982); *J. Organomet. Chem.*, **290**, 199 (1985).
444. F. Calderoni, F. Demartin, M.C. Iapalocci, F. Laschi and P. Zanello, *Inorg. Chem.*, **35**, 898 (1996).
445. D.E. Barber, M. Sabat, E. Sinn and B.A. Averill, *Organometallics,* **14**, 3229 (1995).
446. G. Fischer, G. Sedelmaier, H. Prinzbach, K. Knoll, P. Wilharm, G. Huttner and I. Jibril, *J. Organomet. Chem.*, **297**, 307 (1985).

447. R.D. Pergola, C. Bandini, F. Demartin, E. Diana, L. Garlaschelli, P.L.Stanghellini and P. Zanello, *J. Chem. Soc., Dalton Trans.*, 747 (1996).
448. M.J. Jordan, P.S. White, C.K. Schauer and M.A. Mosley III, *J. Am. Chem. Soc.*, **117**, 5403 (1995).
449. G. Ferguson, R. Hourihane and T.R. Spalding, *Acta Cryst., Sect. C*, **47**, 544 (1991).
450. G.L. Lilley, E. Sinn and B.A. Averill, *Inorg. Chem.*, **25**, 1073 (1986).
451. Mighuey Shieh, Mei-Huey Shieh, Yi-Chou Tsai and Chuen-Her Ueng, *Inorg. Chem.*, **34**, 5088 (1995).
452. H.G. Raubenheimer, . G.J. Kruger and A. Van A. Lombard, *J. Organomet. Chem.*, **323**, 385 (1987).
453. L.J. Arnold, K.M. Mackay and B.K. Nicholson, *J. Organomet. Chem.*, **387**, 197 (1990).
454. A.I. Nekhaev, B.I. Kolobkov, G.G. Aleksandrov, N.A. Parpiev, M.T. Tashev and Kh.B. Dustov, *Izv. Akad. Nauk SSSR, Ser. Khim.*, 458 (1987); *Koord. Khim.*, **14**, 273 (1988); Engl. Ed. p.158.
455. A. Bandyopadhyay, Maoyu Shang, Chang Soo Jun and Th.P. Fehlner, *Inorg. Chem.*, **33**, 3677 (1994).
456. Minghuey Shieh, Puo-Fei Chen, Shie-Ming Peng and Gene-Hsiang Lee, *Inorg. Chem.*, **32**, 3389 (1993).
457. J.T. Jaeger and H. Vahrenkamp, *Organometallics*, **7**, 1746 (1988).
458. R. Grobelny, M. Melnik and J. Mrozinski, *"Cobalt Coordination Compounds", Dolnoslaskie Wydavnictvo Edukacyjne, Wroclaw*, 1996, p.1-493.
459. M. Melnik, T. Šramko, M. Dunaj-Jurèo, A. Sirota, E. Jóna and C.E. Holloway, *Rev. Inorg. Chem.*, **14**, 1 (1994); Supplement ibid, **15**, 139 (1995).
460. M. Melnik, T. Šramko, M. Dunaj-Jurèo and V. Mišút, *Rev. Inorg. Chem.*, **12**, 1 (1992).

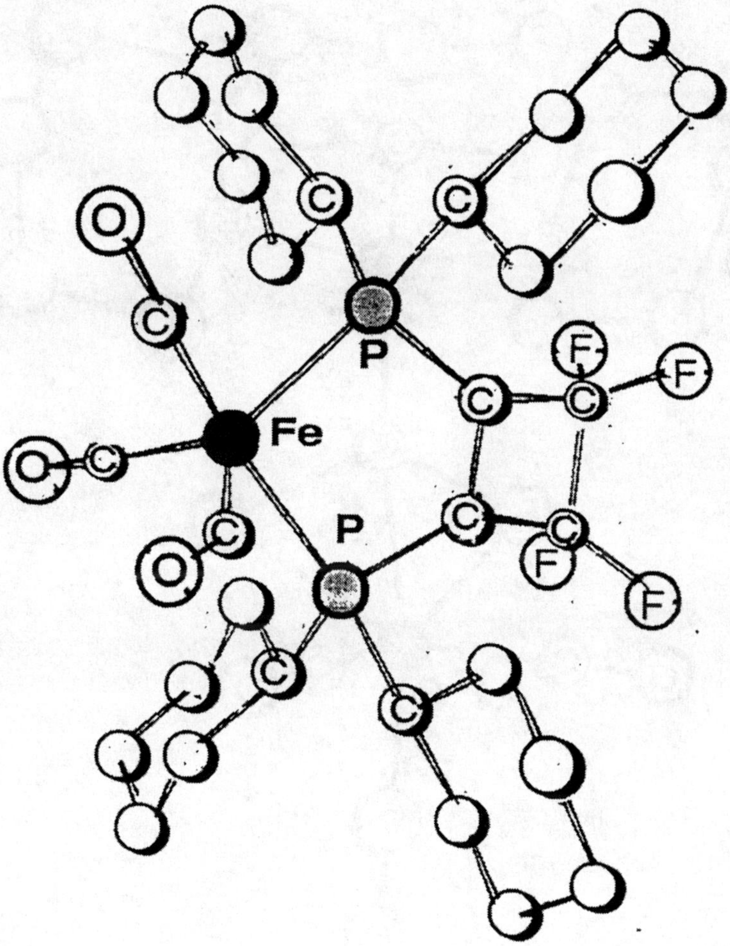
O
C
C
P
F
F
C
C
Fe
O
C
C
C
C
P
C
F
O
F
C
C

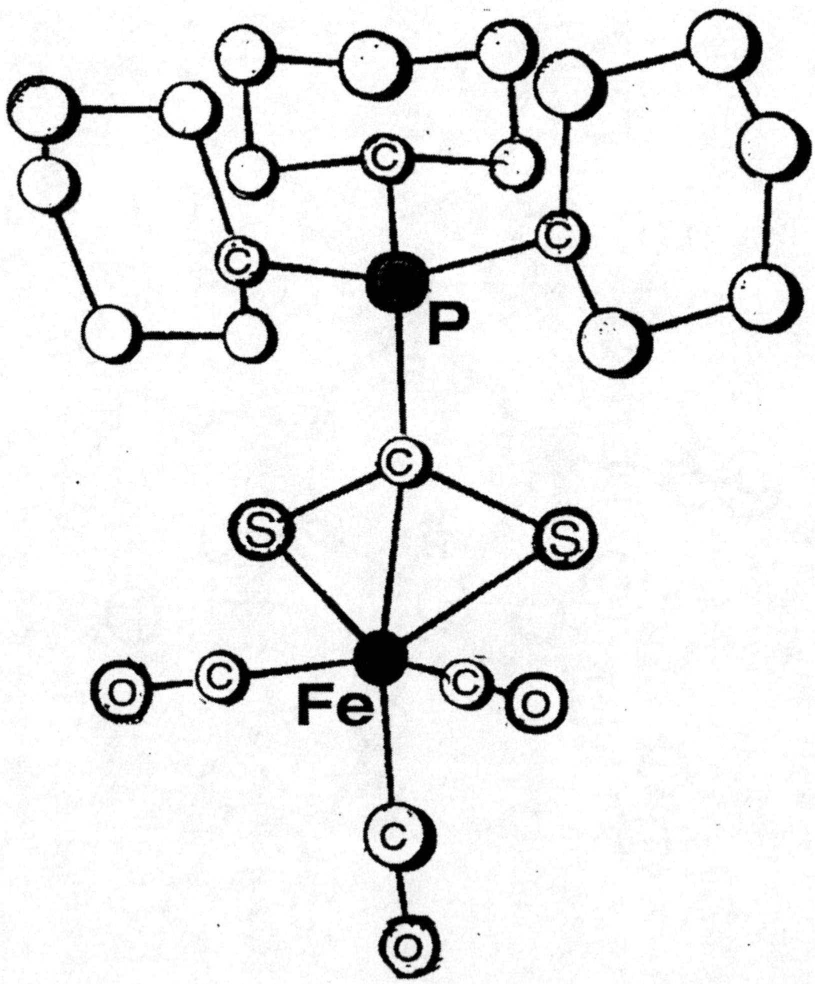
C
C
C
P
C
S
S
Fe
O
C
C
O
C
O

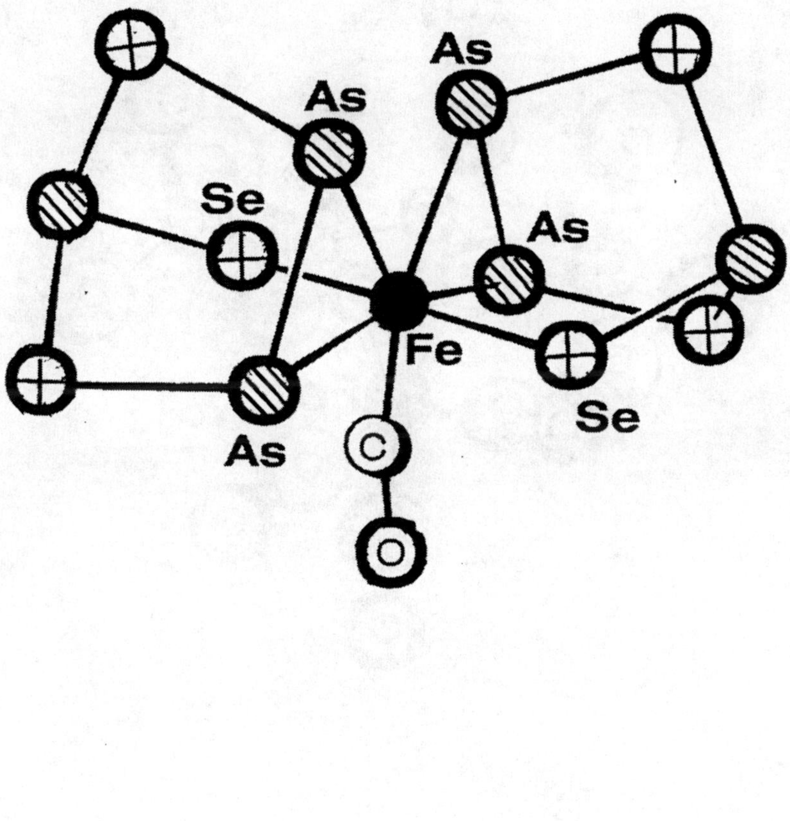
As
As
Se
As
Fe
Se
As
C
O

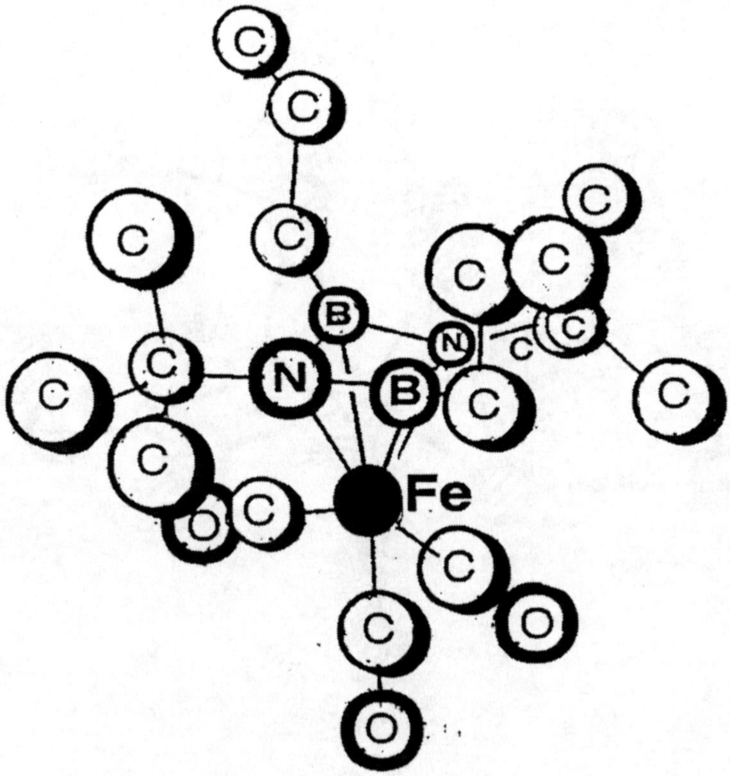
C
C
C
C
C
C
B
N
N
B
C
C
C
C
C
C
C
Fe
C
C
O
C
O
C
O

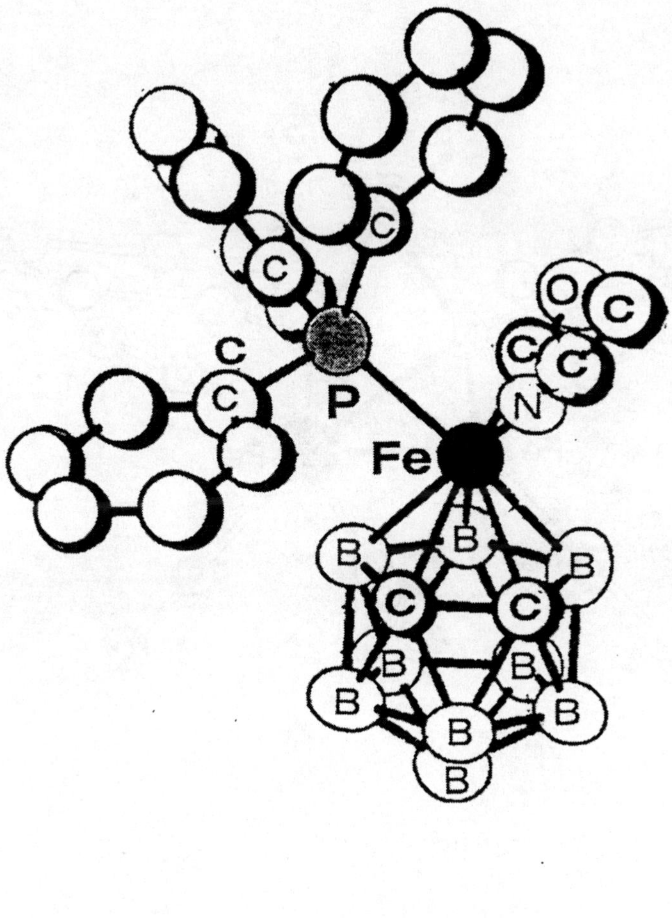
C
C
C
C
P
Fe
O
C
C
C
N
B
B
B
C
C
B
B
B
B
B
B

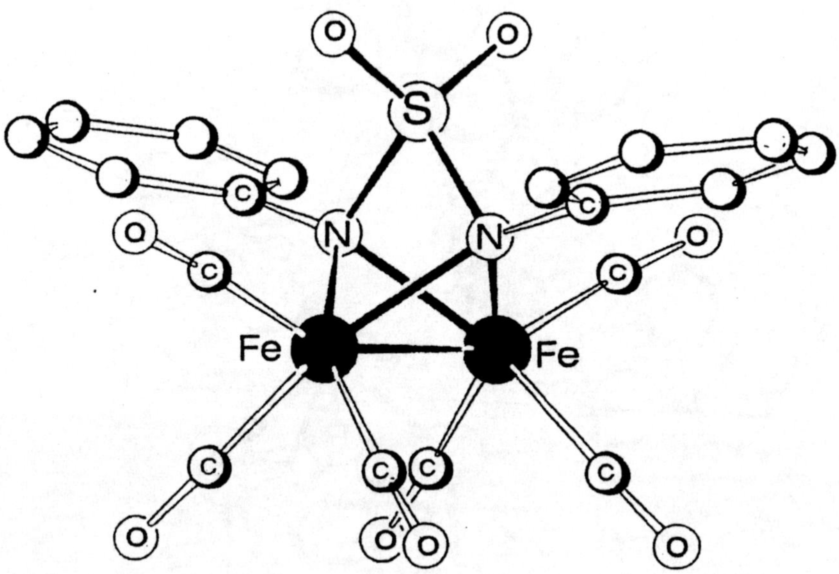
O
O
S
C
C
N
N
O
O
C
C
Fe
Fe
C
C
C
C
O
O
O
O

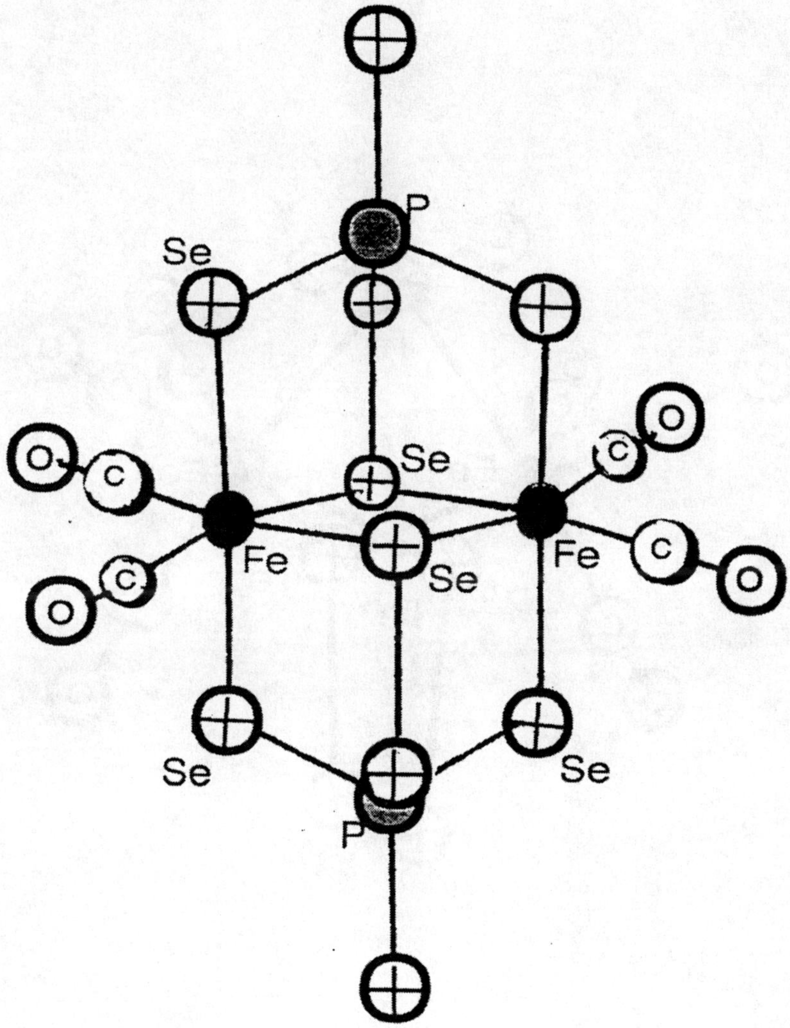

P
Se
Se
O
C
Fe
O
C
Se
Fe
C
O
C
O
Se
Se
P

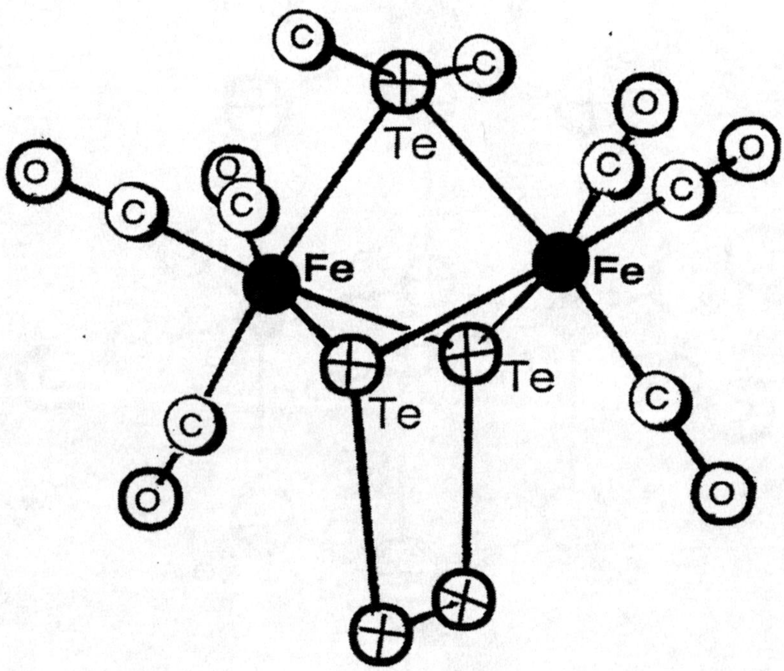
C
C
Te
O
O
O
C
O
C
C
C
Fe
Fe
Te
Te
C
C
O
O

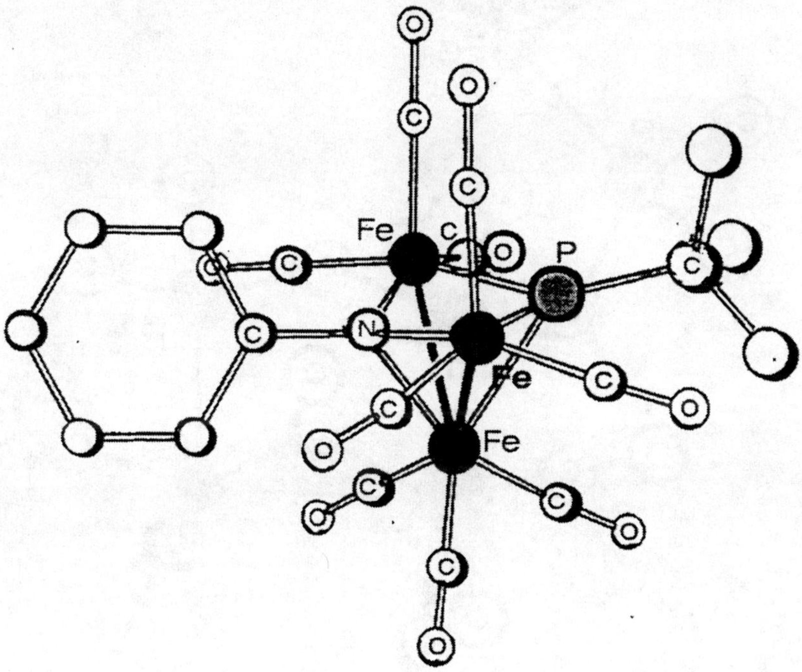
Fe
C
P
N
Fe
Fe

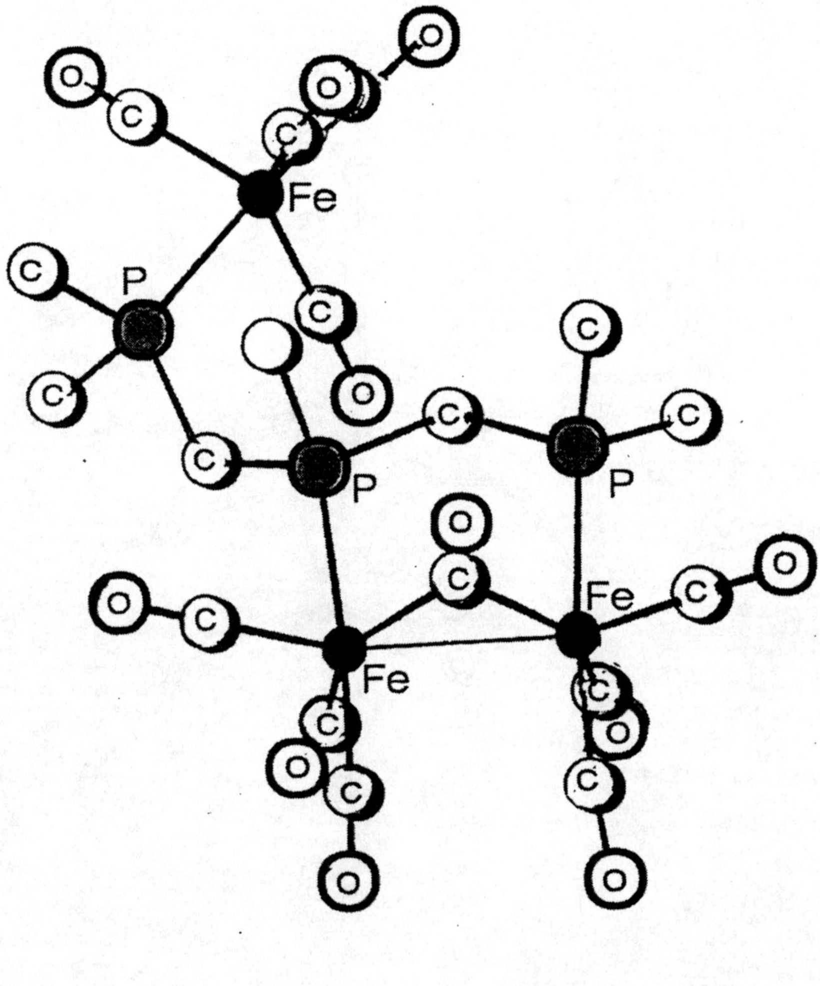
Fe
P
P
P
Fe
Fe

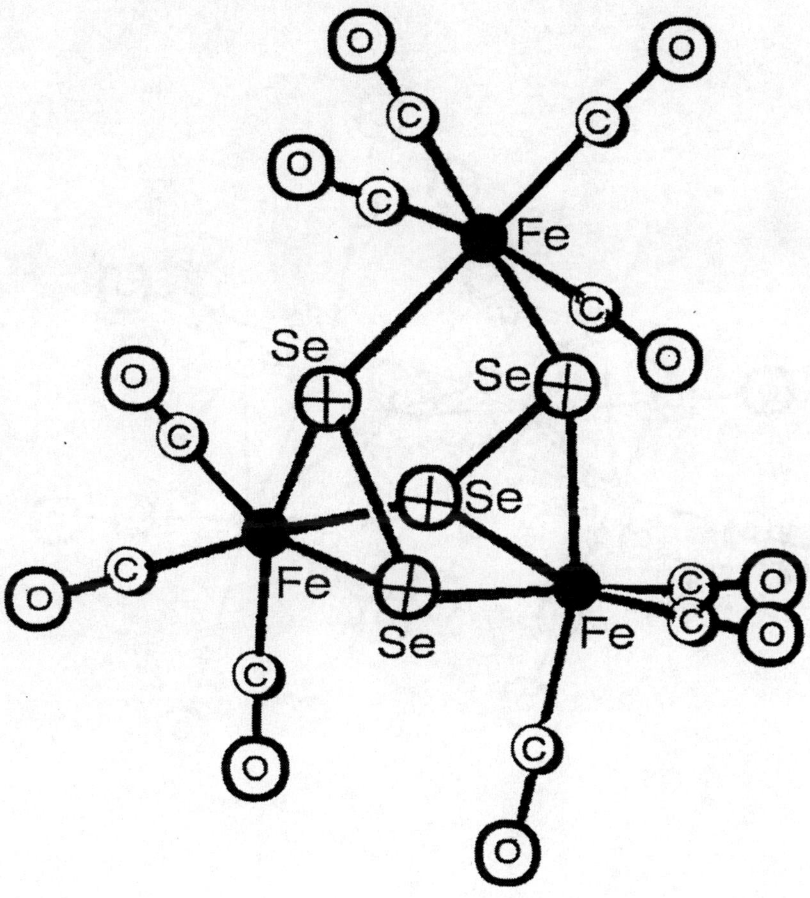
Fe
Se
Se
Se
Fe
Se
Fe

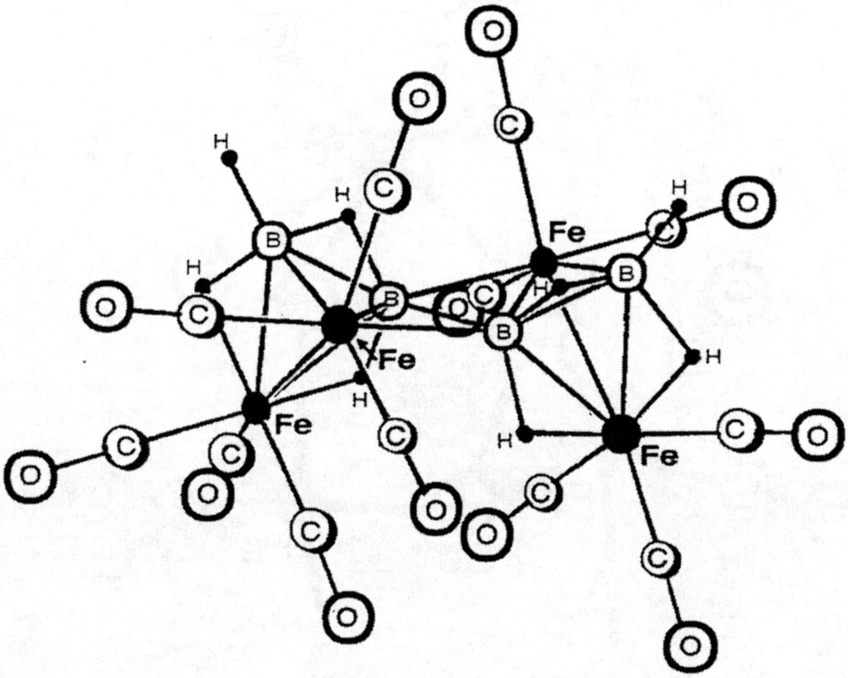

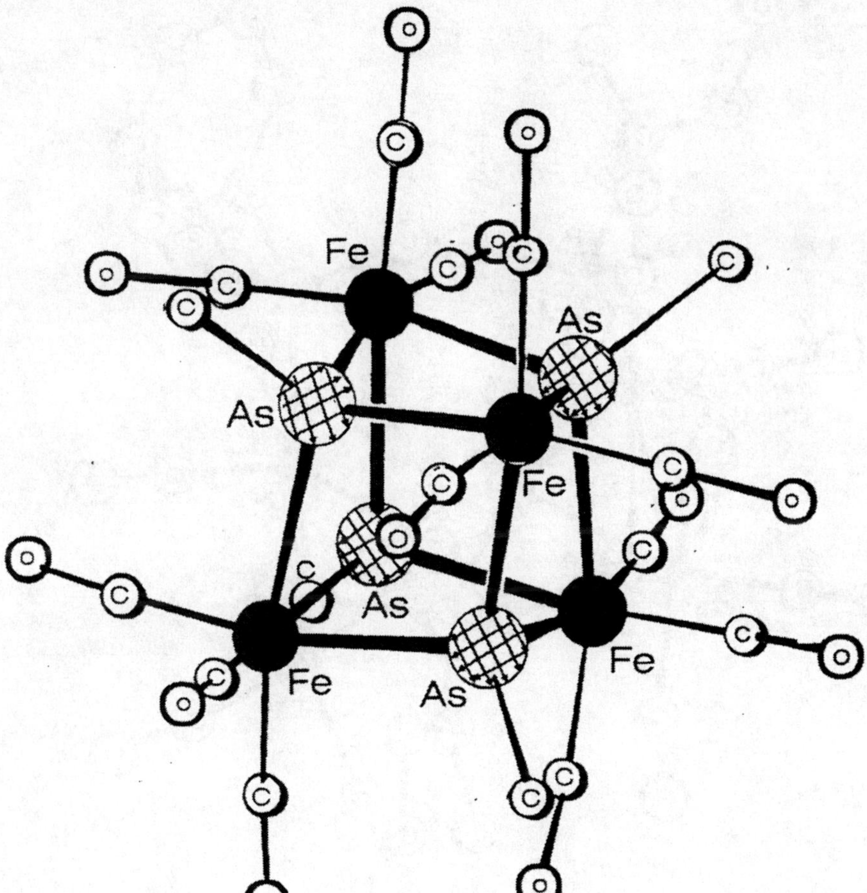
Fe
As
As
Fe
As
Fe
As
Fe

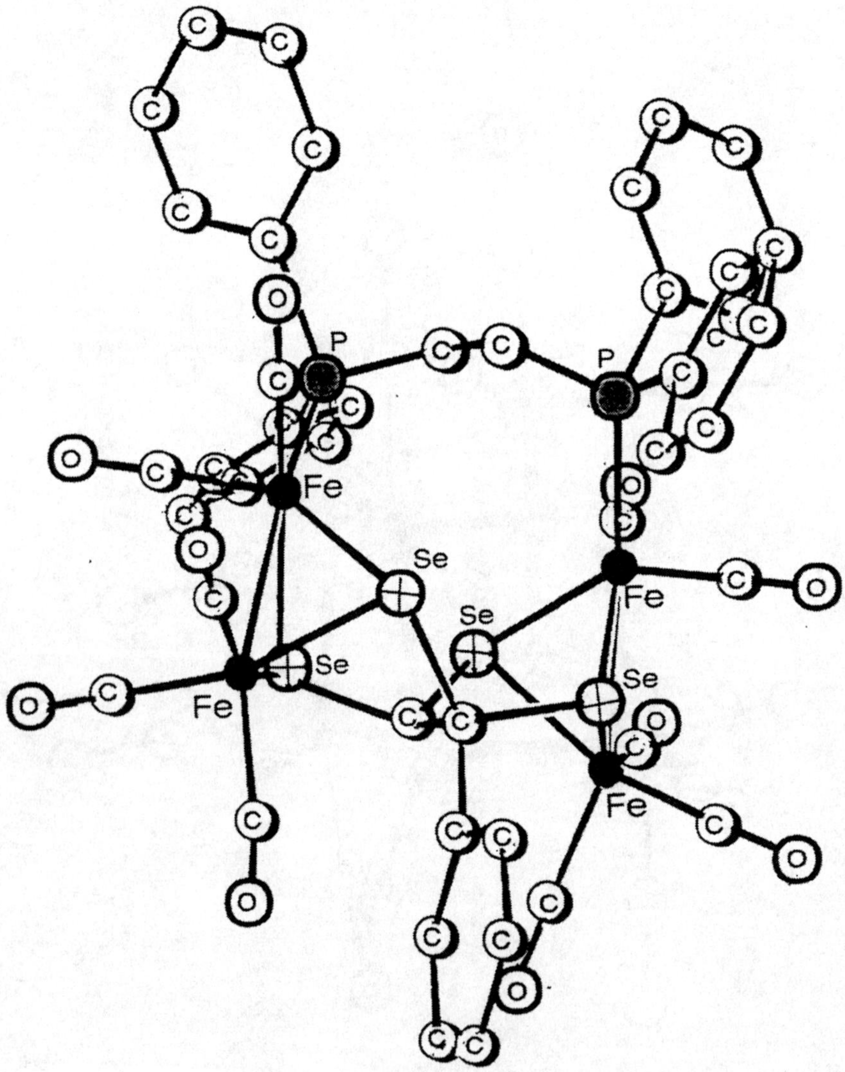
P
P
Fe
Se
Fe
Se
Se
Fe
Se
Fe

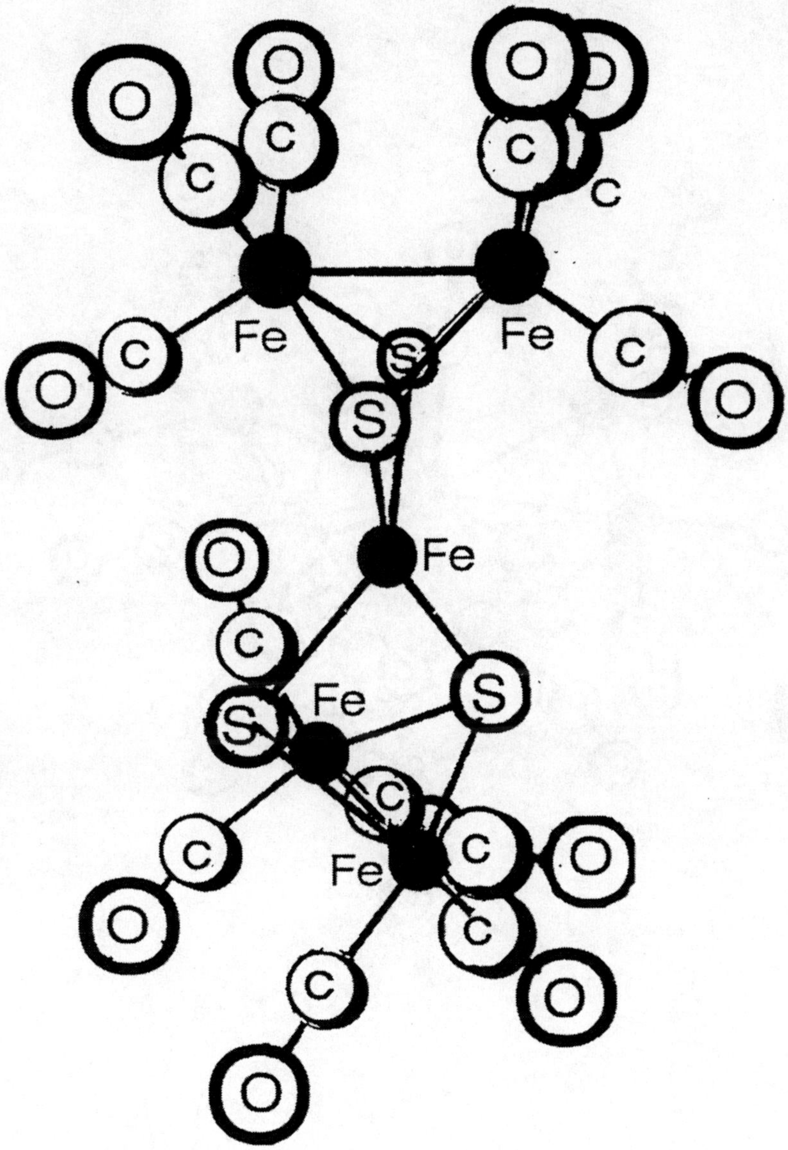
O
O
O
O
C
C
C
C
Fe
Fe
C
C
O
S
O
S
Fe
O
C
S
Fe
S
C
Fe
C
O
O
C
O
C
O

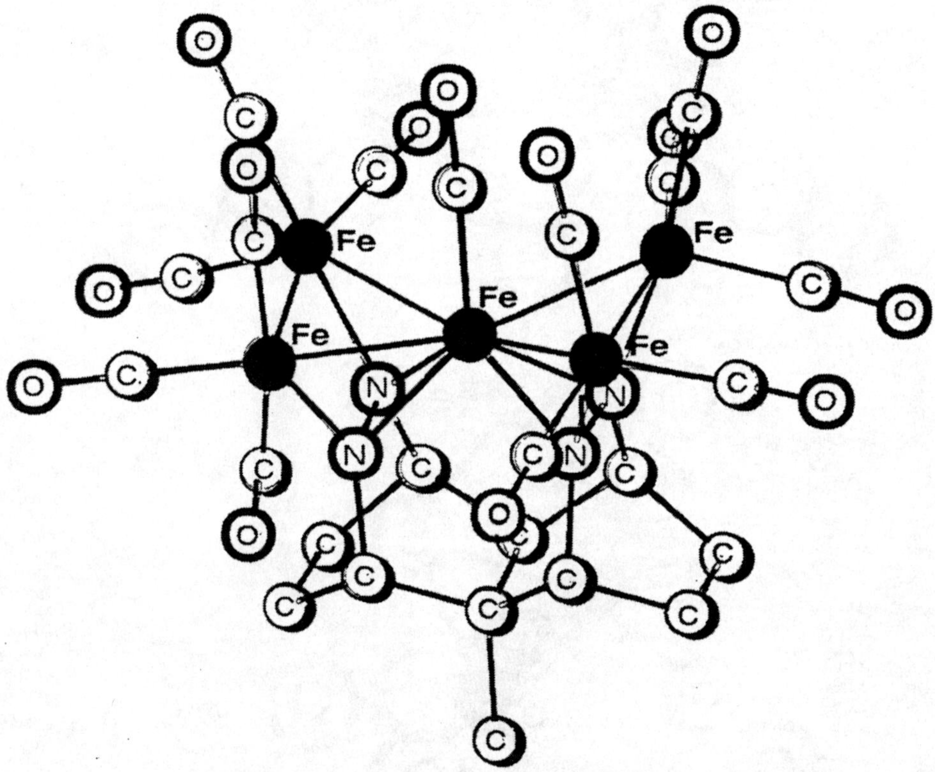
Fe
Fe
Fe
Fe
Fe

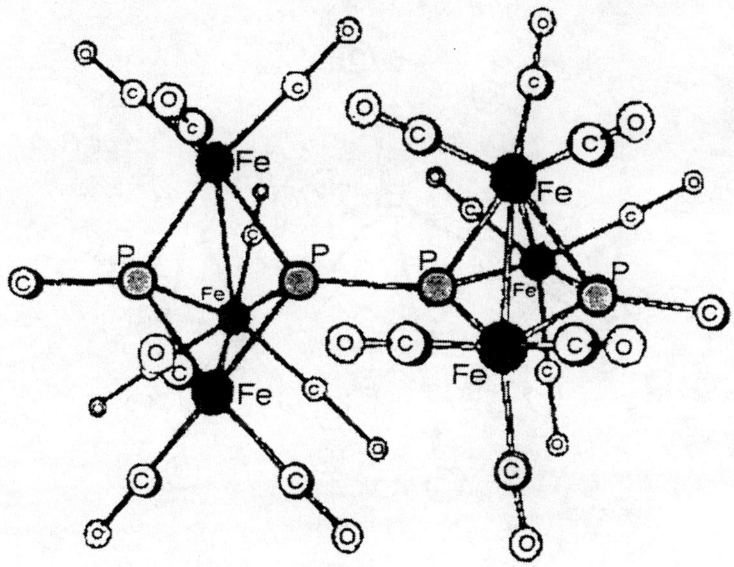
Fe
Fe
Fe
Fe
Fe
Fe
P
P
P
P

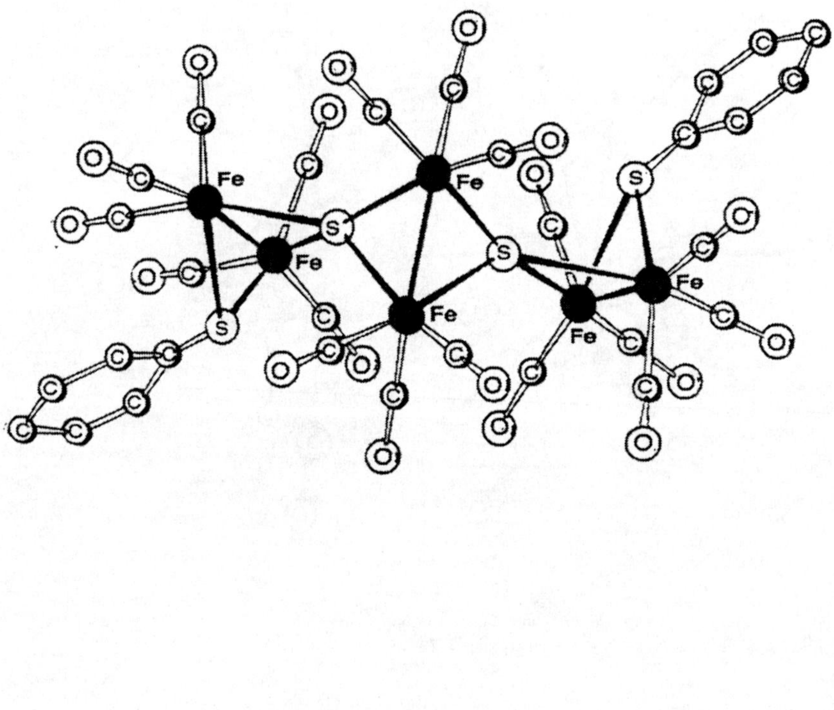
Fe
Fe
Fe
Fe
Fe
Fe
S
S
S
S

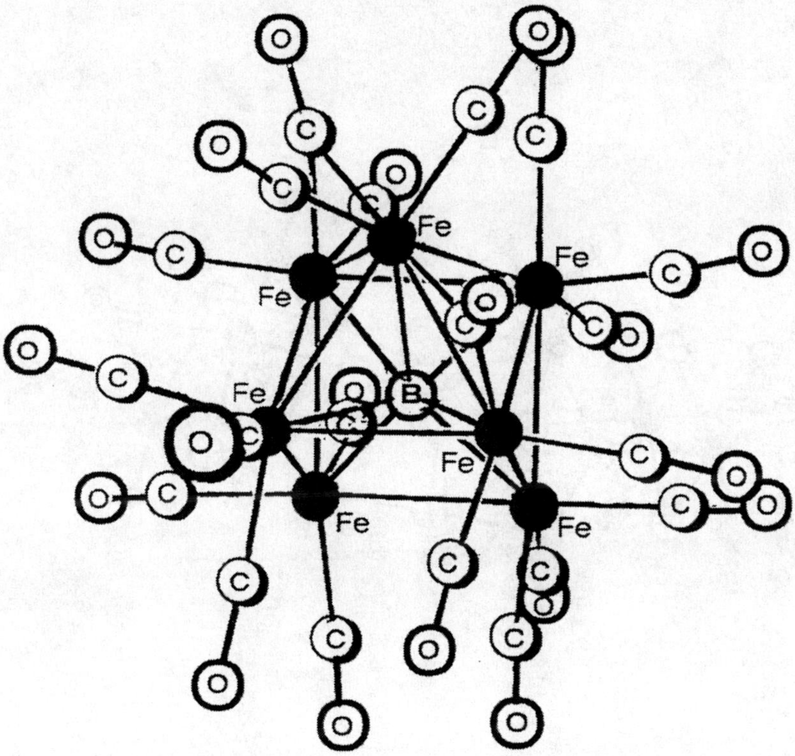
Fe
Fe
Fe
Fe
B
Fe
Fe
Fe

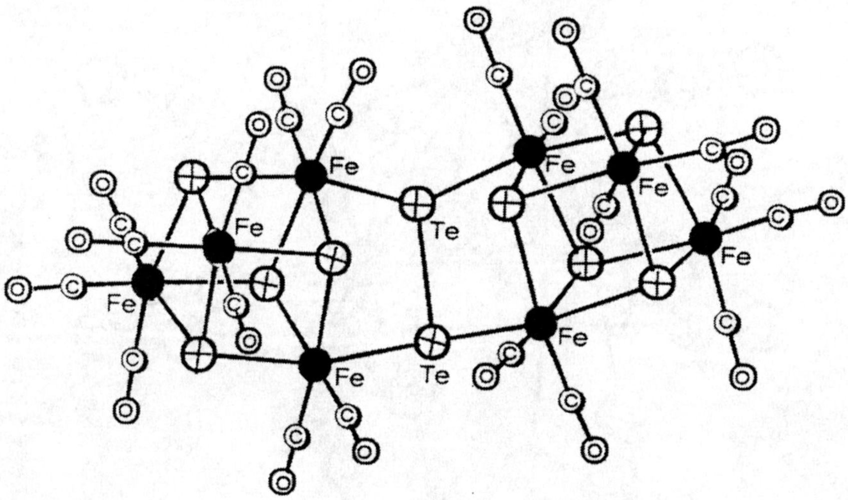
Fe
Fe
Fe
Te
Te
Fe
Fe
Fe
Fe
Fe

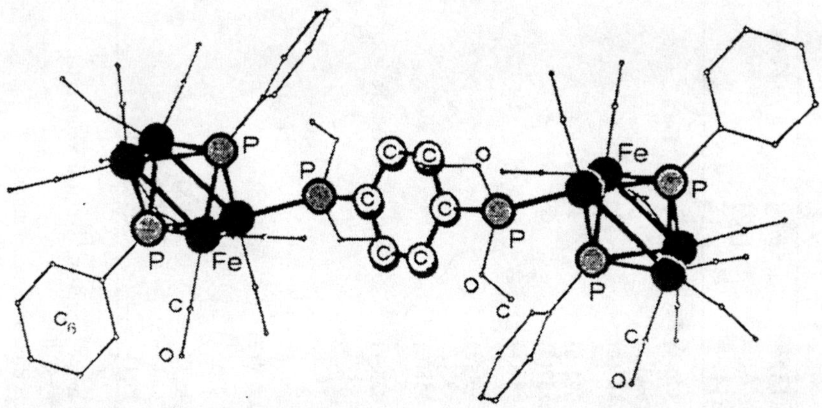
P
P
Fe
C
O
C6
C
C
C
C
C
C
O
P
O
C
Fe
P
P
C
O

Table 1. Crystallographic and Structural Data for Iron Carbonyls with Coordination Number Four[a]

COMPOUND (colour)	CRYST. CL. SPACE G. Z	a [Å] b [Å] c [Å]	α [°] β [°] γ [°]	CHROMO-PHORE	Fe-L [Å]		L-Fe-L [°]	REF.
$Na_2[Fe^{-II}(CO)_4]\cdot 1.5dio$ (pale yellow)	tg $P4_2/m$ 4	10.690(5) 10.690(5) 12.283(6)		FeC_4	C[b] 1.742(6,4)	C,C[b]	105.1(3,9) 129.7(2)	3
$K_2[Fe^{-II}(CO)_4]$ (light gold)	or Fddd 8	23.072(6) 11.557(2) 5.543(1)		FeC_4	C 1.746(4,0)	C,C	101.0-121.0(2)	4
$[Na(crypt)]_2[Fe^{-II}(CO)_4]$ (colourless)	or $P2_12_12_1$ 4	16.360(3) 12.592(2) 23.058(6)		FeC_4	C 1.76(2,5)	C,C	109.5(9,2.7)	4
$Tl[Fe^{-II}(CO)_3(NO)]$ (dark red) (at 240 K)	c $P2_13$ 4	8.853(1)		FeC_3N	C 1.800(8,0) N 1.659(11)	C,C C,N	101.4(3) 116.7(3)	5
$(ppn)[Fe^{-II}(CO)_3(NO)]$ (not given)	m $P2_1/c$ 4	17.533(6) 9.202(17) 22.026(11)	 96.15(3)	FeC_3N	C/N 1.714(9,26)	C/N,C/N	106.3(4)- 113.3(4)	6
$[Fe^{-II}(NO)_2(CO)(PPh_3)]$ (brown)	tr P-1 2	10.96(1) 10.20(1) 10.45(1)	115.84(8) 117.33(8) 78.90(8)	FeN_2CP	C/N 1.709(8,23) P 2.260(3)	C/N,C/N C/N,P	114.4(4,2.9) 103.9(4,6)	7

a Where more than one chemically equivalent distance or angle is present, the mean value is tabulated. The first number in parenthesis is the e.s.d., and the second is the maximum deviation from the mean.

b The chemical indentity of coordinated atom/ligand is specified in these columns.

Table 2. Crystallographic and Structural Data for Iron Carbonyls with Coordination Number Five[a]

COMPOUND (colour)	CRYST. CL. SPACE G. Z	a [Å] b [Å] c [Å]	α [°] β [°] γ [°]	CHROMO-PHORE	Fe-L [Å]	L-Fe-L [°]	REF.
$[Fe^0(CO)_5]$ (not given)	m C2/c 4	not given	not given	FeC_5	OC_{eq}[b] 1.79(3,3) OC_{ax} 1.81(2,0)	C_{eq},C_{eq}[b] 120(1,4) C_{eq},C_{ax} 90(1,2) C_{ax},C_{ax} 176(2)	8
$[Fe^0(CO)_5]$ (not given) (at 200 K)	m C2/c 4	11.807(1) 6.821(2) 9.367(4)	 107.72(1)	FeC_5	OC not given	C,C not given	9
$[Fe^0(CO)_5]$ (not given) (at 198 K)	m C2/c 4	11.740(2) 6.806(1) 9.341(3)	 107.59(1)	FeC_5	OC_{eq} 1.803(3,2) OC_{ax} 1.811(2,0)	C_{eq},C_{eq} 120.00(10,2.22) C_{eq},C_{ax} 89.93(7,46) C_{ax},C_{ax} 178.94(10)	10
$(ppn)[Fe^0(CO)_4CN]$ (gold)	m $P2_1/c$ 4	10.749(4) 12.909(6) 26.801(9)	 97.98(5)	FeC_5	OC_{eq} 1.768(8,14) OC_{ax} 1.723(8) NC_{ax} 1.928(8)	C_{eq},C_{eq} 119.8(4,5.3) C_{eq},C_{ax} 87.4(3,2.6) C_{eq},NC_{ax} 92.7(4,2.4) C_{ax},NC_{ax} 175.9(3)	11
$[C(NMe_2)_3][Fe^0(CO)_4$- $C(O)NMe_2]$ (not given)	tr P-1 2	7.876(3) 10.098(3) 12.067(5)	94.25(3) 90.94(3) 97.30(3)	FeC_5	OC_{eq} 1.774(5,12) OC_{ax} 1.774(5) NC_{ax} 2.053(4)	C_{eq},C_{eq} 119.7(2,10.1) C_{eq},C_{ax} 93.3(2,1.6) C_{eq},NC_{ax} 86.9(2,1.7) C_{ax},NC_{ax} 179.2(5)	12
$[Fe^0(CO)_3(MeCN)_2]$ (yellow)	m $P2_1/c$ 4	12.451(6) 6.564(3) 12.087(6)	 92.45(3)	FeC_5	OC_{eq} 1.785(9,10) $MeNC_{ax}$ 1.867(8,7)	C_{eq},C_{ax} 90.0(3,2.2) C_{ax},C_{ax} 177.1(3)	13
$(ppn)[HFe^0(CO)_4]$ (yellow)	m $P2_1/c$ 4	17.701(9) 9.136(4) 22.251(12)	 95.87(4)	FeC_4H	OC_{eq} 1.75(2,4) OC_{ax} 1.72(2) H_{ax} 1.57(12)	C_{eq},C_{eq} 117.6(9,4.8) C_{eq},C_{ax} 99.1(8,2.8) C_{eq},H_{ax} 81(5,3) C_{ax},H_{ax} 175(5)	14
$[Fe^{II}(im)_6][HFe^0(CO)_4]_2$ (not given)	tr P-1 1	10.61(1) 10.61(1) 10.61(1)	84.14(9) 84.15(9) 84.19(9)	FeC_4H	not given	not given	15

Table 2. Continued 2

$Fe^{II}(im)_6][HFe^0(CO)_4]_2$ (not given) (at 225 K)	tr P-1 1	10.49(6) 10.49(1) 10.49(1)	84.26(6) 84.26(6) 84.26(7)	FeC_4H	not given	not given	
$[Fe^0(CO)_4py]$ (orange)	m $P2_1/n$ 4	8.198(6) 10.488(5) 12.080(8)	98.74(6)	FeC_4N	OC_{eq} 1.805(8,6) OC_{ax} 1.772(7) pyN_{ax} 2.046(5)	C_{eq},C_{eq} 120.0(3,2.9) C_{eq},C_{ax} 89.8(3,9) C_{eq},N_{ax} 98.3(6,2.2) C_{ax},N_{ax} 176.2(2)	16
$[Fe^0(CO)_4pz]$ (orange)	m $P2_1/c$ 4	7.760(2) 10.593(3) 12.801(3)	105.69(2)	FeC_4N	OC_{eq} 1.810(4,6) OC_{ax} 1.774(4) pzN_{ax} 2.031(2)	C_{eq},C_{eq} 120.0(2,5.3) C_{eq},C_{ax} 89.4(2,7) C_{eq},N_{ax} 90.7(1,3.8) C_{ax},N_{ax} 175.8(2)	16
$[Fe^0(CO)_4\{Ph(CH)_3NMe\}]$ (red)	m $P2_1/n$ 4	7.192(4) 17.71(1) 11.355(8)	90.39(5)	FeC_4N	OC_{eq} 1.790(7,17) OC_{ax} 1.744(6) N_{ax} 2.015(4)	C_{eq},C_{eq} 120.0(3,9.1) C_{eq},C_{ax} 89.1(3,1.8) C_{eq},N_{ax} 91.1(2,3.6) C_{ax},N_{ax} 174.5(2)	17
$[Fe^0(CO)_4(mcpp)]$ (not given)	m $P2_1/c$ 4	10.113(2) 27.548(7) 6.938(2)	99.53(2)	FeC_4N	OC_{eq} 1.807(4,15) OC_{ax} 1.773(4) N_{ax} 1.979(2)	C_{eq},C_{eq} 120.0(2,1.7) C_{eq},C_{ax} 89.3(2,1.1) C_{eq},N_{ax} 90.7(1,1) C_{ax},N_{ax} 178.7(1)	18
$[Fe^0(CO)_4(pcp)]$ (orange)	m $P2_1/c$ 4	9.208(2) 11.999(2) 23.911(6)	94.50(2)	FeC_4P	OC_{eq} 1.780(5,13) OC_{ax} 1.793(5) P_{ax} 2.225(1)	C_{eq},C_{eq} 115.7(2) C_{eq},P_{ax} 86.6(1) 123.8(1) C_{ax},P_{ax} not given	19
$[Fe^{II}(CO)([14]annulene)]\cdot0.5tol$ (not given)	tr P-1 2	8.6254(19) 13.0137(31) 10.9199(32)	102.32(2) 104.39(2) 78.60(2)	FeN_4C	N_{eq} 1.927(4,8) OC_{ax} 1.694(4)	N_{eq},N_{eq} 61.3(1) 88.7(1,6.6) 164.0(1) N_{eq},C_{ax} 98.7(1,2.7)	20
$K[Fe^{II}(CO)(Pr^i\text{-cma})]\cdot dmf$ (deep blue)	tr P-1 not given	9.4318(1) 10.9041(3) 12.8403(3)	82.938(1) 87.196(1) 83.674(1)	FeN_4C	N_{eq} 1.995(3,5) N_{ax} 1.991(3) OC_{ax} 1.749(3)	N_{eq},N_{eq} 117.9(1,2.1)	21

Table 2. Continued 3

Compound	Cryst. cl. / Space group / Z	a, b, c	α, β, γ	Chromophore	Fe–L	L–Fe–L	Ref.
$[Fe^{0}(CO)_4(PBu^t_3)]$ (not given)	or $P2_12_12_1$ 4	18.842(7) 10.062(4) 9.790(2)		FeC_4P	OC_{eq} 1.792(5,13) OC_{ax} 1.768(4) P_{ax} 2.364(1)	C_{eq},C_{eq} 119.7(2,4.5) C_{eq},C_{ax} 86.8(2,2.8) C_{eq},P_{ax} 93.3(1,1.4) C_{ax},P_{ax} 175.7(1)	22
$[Fe^{0}(CO)_4PHPh_2]$ (honey-coloured)	m $P2_1/c$ 4	11.424(4) 12.459(7) 11.644(7)	105.370(35)	FeC_4P	OC_{eq} 1.793(9,10) OC_{ax} 1.792(8) P_{ax} 2.237(2)	C_{eq},C_{eq} 120.0(4,4.1) C_{eq},C_{ax} 90.9(4,1.4) C_{eq},P_{ax} 89.2(2,1.5) C_{ax},P_{ax} 177.1(3)	23
$[Fe^{0}(CO)_4(PPh_3)]$ (yellow) (at 238 K)	tr P-1 2	10.348(2) 10.709(7) 9.809(6)	112.73(5) 94.39(2) 90.12(2)	FeC_4P	OC_{eq} 1.795(4,3) OC_{ax} 1.795(4) P_{ax} 2.244(1)	C_{eq},C_{eq} 119.9(2,3.7) C_{eq},C_{ax} 91.2(2,1.1) C_{eq},P_{ax} 88.8(1,6) C_{ax},P_{ax} 178.3(1)	24
$[Fe^{0}(CO)_4\{P(o\text{-tolyl})_3\}]$ (white)	m $P2_1/n$ 4	10.188(3) 10.429(2) 21.755(6)	99.79(2)	FeC_4P	OC_{eq} 1.763(5,0) OC_{ax} 1.777(6) P_{ax} 2.306(1)	C_{eq},C_{eq} 120.0(2) C_{eq},C_{ax} 89.4(3) C_{eq},P_{ax} 90.6(2)	25, 26
$[Fe^{0}(CO)_4\{P(o\text{-tolyl})_2CH_2Ph\}]$ (white)	tr P-1 4	9.618(3) 15.282(3) 17.130(5)	66.92(2) 79.51(2) 86.42(2)	FeC_4P	OC_{eq} 1.783(7,0) OC_{ax} 1.780(7) P_{ax} 2.271(2)	C_{eq},C_{eq} 120.0(3,9.2) C_{eq},C_{ax} 89.9(3) C_{eq},P_{ax} 90.2(2) C_{ax},P_{ax} 176.3(2)	26
$[Fe^{0}(CO)_4\{\eta^1\text{-}(o\text{-tolyl})_2\text{-}PP(o\text{-tolyl})_2\}]$ (white)	or Pbca 8	15.947(6) 20.009(3) 15.470(4)		FeC_4P	OC_{eq} 1.78(1,0) P_{eq} 2.301(2) OC_{ax} 1.78(1,0)	C_{eq},C_{eq} 131.9(4) C_{eq},C_{ax} 87.2(4) C_{ax},C_{ax} 165.7(4) C_{eq},P_{eq} 114.0(3,5.5) C_{ax},P_{eq} 97.1(3,4.9)	26
$[Fe^{0}(CO)_4(dppm)]$ (not given)	m $P2_1/c$ 4	8.698(3) 27.478(3) 11.161(2)	95.77(2)	FeC_4P	OC_{eq} 1.79(1,1) OC_{ax} 1.79(1) P_{ax} 2.267(3)	C_{eq},C_{eq} 120.0(5,5.7) C_{eq},C_{ax} 90.1(5,2.5) C_{eq},P_{ax} 89.9(4,2.5) C_{ax},P_{ax} 176.9(4)	27
$[Fe^{0}(CO)_4(dppe)]$ (not given)	m $P2_1/c$ not given	8.518(3) 21.897(6) 14.341(3)	93.74(3)	FeC_4P	OC_{eq} 1.790(6,2) OC_{ax} 1.771(6) P_{ax} 2.246(2)	C_{eq},C_{eq} 120.0(3,1.4) C_{eq},C_{ax} 91.5(3,1.1) C_{eq},P_{ax} 88.5(2,5) C_{ax},P_{ax} 179.3(2)	28

Table 2. Continued 4

$[Fe^{0}(CO)_4\{(BNPr^{i}_2)_3P_2\}]$ (not given)	or Cmca 8	15.949(5) 15.473(4) 27.207(9)		FeC_4P	OC_{eq} 1.765(15,0) OC_{ax} 1.749(13) P_{ax} 2.260(3)	not given	29
$[Fe^{0}(CO)_4\{(2,4,6\text{-}Bu^{t}_3C_6H_2)_2P_2\}]$ (dark red)	m $P2_1/c$ 4	21.251(6) 9.793(2) 20.99(2)	108.73(6)	FeC_4P	OC_{eq} 1.794(6,9) P_{eq} 2.215(1) Oc_{ax} 1.797(6,9)	C_{eq},C_{eq} 117.1(2) C_{eq},P_{eq} 121.4(2,4.3) C_{eq},C_{ax} 88.9(3,2.8) C_{ax},P_{eq} 92.3(2,3.5) C_{ax},C_{ax} 174.9(2)	30
$[Fe^{0}(CO)_4(asp)]$ (yellow)	m $P2_1/c$ 4	11.284(4) 10.800(6) 19.755(8)	96.85(2)	FeC_4P	OC_{eq} 1.74(1,3) OC_{ax} 1.74(2) P_{ax} 2.224(3)	C_{eq},C_{eq} 120.0(6,3.9) C_{eq},C_{ax} 90.6(6,4) C_{eq},P_{ax} 89.4(4,1.6) C_{ax},P_{ax} 178.8(9)	31
$[Fe^{I}(CO)_4\{HB(pz)_3\}]$ (not given)	m $P2_1$ 2	9.590(1) 11.611(3) 10.694(1)	93.41(1)	FeC_4P	OC_{eq} not given OC_{ax} not given P_{ax} 2.274(2)	not given	32
$[Fe^{0}(CO)_4\{(CF_3)_2PNPPh_3\}]$ (not given)	tr P-1 2	9.215(2) 10.758(4) 14.555(6)	111.290(0) 91.920(0) 97.750(10)	FeC_4P	OC_{eq} 1.749(12,0) OC_{ax} 1.771(10) P_{ax} 2.184(2)	C_{eq},C_{eq} 122.8(5) C_{eq},C_{ax} 92.7(4,1.1) C_{ax},P_{ax} 176.4(5)	33
$[Fe^{0}(CO)_4\{PPh(PPh_2)_2\}]$ (not given)	tr P-1 2	12.095(4) 14.284(6) 9.769(3)	111.01(3) 95.28(3) 98.21(4)	FeC_4P	OC_{eq} 1.780(6,5) P_{eq} 2.262(1) OC_{ax} 1.778(6,13)	C_{eq},C_{eq} 125.7(2) C_{eq},P_{eq} 117.2(2,8) C_{eq},C_{ax} 88.5(3,1.9) C_{ax},P_{eq} 93.1(2,4) C_{ax},C_{ax} 173.5(2)	34
$[Fe^{0}(CO)_4\{P_2(Pr^{i}_2NB)_2(SiPh_2)\}]$ (orange)	m $P2_1/n$ 4	14.403(3) 15.550(4) 15.632(3)	102.07(1)	FeC_4P	OC_{eq} 1.782(5,2) OC_{ax} 1.782(5) P_{ax} 2.283(1)	not given	35
$[Fe^{0}(CO)_4\{P_2(tmpB)_2(SiMe_2)\}]^{c}$ (yellow)	m $P2_1/c$ 8	11.132(2) 32.351(9) 17.504(4)	92.19(2)	FeC_4P	OC_{eq} 1.772(15,16) OC_{ax} 1.772(15) P_{ax} 2.328(3)	not given	35
				FeC_4P	OC_{eq} 1.771(16,10) OC_{ax} 1.771(16) P_{ax} 2.327(4)	not given	

Table 2. Continued 5

$[Fe^{0}(CO)_4\{P(SiMe_3)_3\}]$ (orange)	m Cc 4	9.548(2) 10.052(3) 15.552(6)	 98.12(2)	FeC_4P	OC_{eq} OC_{ax} P_{ax}	1.79(2,5) 1.74(2) 2.338(4)	C_{eq},C_{eq} 120.0(9,5.2) C_{eq},C_{ax} 89.9(8,3.7) C_{eq},P_{ax} 90.3(7,1.4) C_{ax},P_{ax} 174.9(6)	36
$[Fe^{0}(CO)_4\{P(Me_2N)_3\}]$ (yellow)	m $P2_1/c$ 4	9.606(1) 9.508(5) 18.335(2)	 107.87(1)	FeC_4P	OC_{eq} OC_{ax} P_{ax}	1.787(5,6) 1.793(6) 2.245(1)	C_{eq},C_{eq} 120.0(2,2.7) C_{eq},C_{ax} 90.9(2,1.0) C_{eq},P_{ax} 89.1(2,8) C_{ax},P_{ax} 178.1(2)	37
$[Fe^{0}(CO)_4\{P(Et_2N)_2\}][AlCl_4]$ (orange) (at 165 K)	m $P2_1/c$ 16	17.547(8) 29.007(20) 17.876(5)	 91.80(3)	FeC_4P	OC_{eq} P_{eq} OC_{ax}	1.82(4,1) 2.10(5) 1.78(4,2)	C_{eq},C_{eq} 114(3) C_{ax},C_{ax} 178(1)	38
$[Fe^{0}(CO)_4(fmap)]$ (yellow)	m $P2_1/c$ 4	6.6530(5) 16.3472(16) 11.5939(8)	92.777(7)	FeC_4P	OC_{eq} OC_{ax} P_{ax}	1.796(6,3) 1.803(6) 2.174(1)	C_{eq},C_{eq} 120.0(3,6.3) C_{eq},C_{ax} 91.0(3,2.3) C_{eq},P_{ax} 89.1(2,2.9) C_{ax},P_{ax} 175.4(2)	39
$[Fe^{0}(CO)_4(\eta^{1}\text{-tsamp})]$ (dark orange)	tr P-1 2	10.567(3) 15.129(7) 10.562(3)	97.44(3) 115.50(2) 106.08(3)	FeC_4P	OC_{eq} P_{eq} OC_{ax}	1.781(7,13) 2.208(2) 1.791(4,4)	C_{eq},C_{eq} 110.4(3) C_{eq},P_{eq} 124.8(3,7) CP_{eq},C_{ax} 90.8(3,4.0) C_{ax},C_{ax} 173.1(4)	40
$[Fe^{0}(CO)_4(Pr^{i}_2NPO)_4]$ (not given)				FeC_4P	OC_{eq} OC_{ax} P_{ax}	1.791(9,30) 1.763(10) 2.189(2)	C_{eq},C_{eq} 107.8(3,2.4) 144.5(3) C_{eq},C_{ax} 91.1(4,6.0) C_{eq},P_{ax} 91.3(3,10.3) C_{ax},P_{ax} 161.3(2)	41
$[Fe^{0}(CO)_4\{PC_4H_4Me_2)Ge\text{-}(cp')Me_2\}]$ (not given) (at 105 K)	m $P2_1/n$ 4	9.677(3) 12.9870(16) 18.327(3)	 95.810(19)	FeC_4P	OC_{eq} OC_{ax} P_{ax}	1.778(6,7) 1.777(6) 2.251(2)	C_{eq},C_{eq} 120.0(2,8.4) C_{eq},C_{ax} 91.5(3,2.0) C_{eq},P_{ax} 88.7(2,1.7) C_{ax},P_{ax} 176.3(2)	42
$[Fe^{0}(CO)_3(CNPr^{i}_2)PPh_3]BCl_4$ (yellow orange) (at 233 K)	m $P2_1/n$ 4	13.120(2) 17.997(3) 14.683(3)	 92.98(1)	FeC_4P	OC_{eq} NC_{eq} OC_{ax} P_{ax}	1.823(6,4) 1.734(6) 1.806(6) 2.264(2)	NC_{eq},P_{ax} 88.6(2)	43

Table 2. Continued 6

Compound	Crystal data	a, b, c	Angle	Chromophore	Fe–L		L–Fe–L		Ref.
$(NEt_4)[trans\text{-}Fe^0(CO)_3(CHO)\text{-}$ $(Me_2PhO)_3P]$ (not given)	m C2/c 8	33.240(4) 13.142(3) 16.972(4)	100.02(3)	FeC_4P	OC_{eq} HOC_{ax} P_{ax}	1.760(5,18) 1.949(6) 2.137(1)	C_{eq},C_{eq} C_{eq},C_{ax} C_{eq},P_{ax} C_{ax},P_{ax}	119.6(2,4.7) 86.3(2,3.0) 93.8(1,3.0) 175.4(2)	44
$(NEt_4)[trans\text{-}Fe^0(CO)_3(CHO).$ $\{(OPh)_3P\}]$ (at 168 K)	m $P2_1/n$ 4	10.037(2) 18.588(3) 16.292(4)	90.05(2)	FeC_4P	OC_{eq} HOC_{ax} P_{ax}	1.766(3,10) 1.976(3) 2.124(1)	C_{eq},C_{eq} C_{eq},C_{ax} C_{eq},P_{ax} C_{ax},P_{ax}	119.6(1,3.4) 86.5(1,1.1) 93.5(1,3.8) 176.2(1)	45
$(NEt_4)[trans\text{-}Fe^0(CO)_3(COMe).$ $\{(OPh)_3P\}]$ (light tan) (at 203 K)	m $P2_1/n$ 4	9.956(2) 19.374(4) 16.180(3)	96.35(2)	FeC_4P	OC_{eq} $MeOC_{ax}$ P_{ax}	1.772(4,7) 2.006(4) 2.117(1)	C_{eq},C_{eq} C_{eq},C_{ax} C_{eq},P_{ax} C_{ax},P_{ax}	119.8(2,1.8) 87.4(2,4.8) 92.5(1,4.0) 177.3(1)	45
$trans\text{-}[Fe^0(CO)_3(C_5H_{10}N_2)(PEt_3)]^c$ (cream)	or $P_{ca}2_1$ 8	15.952(12) 7.982(7) 28.444(8)		FeC_4P	OC_{eq} NC_{ax} P_{ax}	1.773(12,27) 1.951(10) 2.202(3)	C_{eq},C_{eq} C_{eq},C_{ax} C_{eq},P_{ax} C_{ax},P_{ax}	120.0(6,7.9) 91.0(5,8.4) 89.3(4,5) 171.5(3)	46
				FeC_4P	OC_{eq} NC_{ax} P_{ax}	1.774(9,30) 1.949(10) 2.214(3)	C_{eq},C_{eq} C_{eq},C_{ax} C_{eq},P_{ax} C_{ax},P_{ax}	119.9(5,11.1) 91.7(5,6.4) 88.7(4,9) 172.8(3)	
$[Fe^0(CO)_4(C_4H_8S_2)]$ (gold)	or $Pn2_1a$ 4	13.488(3) 9.058(2) 9.771(2)		FeC_4S	OC_{eq} OC_{ax} S_{ax}	1.764(9,27) 1.789(9) 2.288(2)	C_{eq},C_{eq} C_{eq},C_{ax} C_{eq},S_{ax} C_{ax},S_{ax}	120.00(54,4.64) 91.73(40,7.09) 89.92(31,3.60) 179.44(28)	47
$(NEt_4)[Fe^0(CO)_4(Cl_3Si)]$ (not given)	m $P2_1$ 2	8.390(1) 10.482(1) 11.528(1)	77.32(2)	FeC_4Si	OC_{eq} OC_{ax} Si_{ax}	1.75(4,6) 1.77(3) 2.224(9)	C_{eq},C_{eq} C_{eq},C_{ax} C_{eq},Si_{ax} C_{ax},Si_{ax}	119.5(1.4,5.1) 94.2(1.7,1.0) 85.8(1.3,1.4) 178.0(1.1)	48
$[Fe^0(CO)_4(sil)]$ (colourless) (at 193 K)	tr P-1 2	8.284(2) 8.404(2) 13.497(3)	97.02(1) 93.68(1) 91.25(1)	FeC_4Si	OC_{eq} OC_{ax} Si_{ax}	1.775(3,4) 1.786(3) 2.259(1)	C_{eq},C_{eq} C_{eq},C_{ax} C_{eq},Si_{ax} C_{ax},Si_{ax}	119.7(1,4.6) 93.3(1,2.3) 86.8(1,3.6) 175.2(1)	49

Table 2. Continued 7

$[Fe^0(CO)_4(MePhsic)]$ (cream)	or Pbca 8	17.416(2) 18.775(4) 18.471(4)		FeC_4Si	OC_{eq} 1.757(3,24) OC_{ax} 1.785(3) Si_{ax} 2.277(1)	not given	50
$[Fe^0(CO)_4(Ph_2sic)]$ (cream)	m $P2_1/n$ 4	18.456(2) 9.797(1) 18.970(1)	91.50(1)	FeC_4Si	OC_{eq} 1.753(6,11) OC_{ax} 1.768(6) Si_{ax} 2.272(1)	not given	50
$[Fe^0(CO)_4\{(OBu^t)_2Si(hmpa)\}]$ (colourless)	m $P2_1/n$ 4	10.069(1) 16.950(1) 15.986(1)	94.17(1)	FeC_4Si	OC_{eq} 1.761(7,8) OC_{ax} 1.758(6) Si_{ax} 2.289(2)	C_{eq},Si_{ax} 86.2(2,1.3) C_{ax},Si_{ax} 177.1(2)	51
$[Fe^0(CO)_4\{Me_2Si(hmpa)\}]$[c] (colourless) (at 233 K)	m $P2_1/c$ 8	8.353(1) 16.079(3) 30.445(4)	106.74(1)	FeC_4Si	OC_{eq} 1.761(7,14) OC_{ax} 1.789(7) Si_{ax} 2.279(2)	not given	52
				FeC_4Si	OC_{eq} 1.749(7,8) OC_{ax} 1.782(7) Si_{ax} 2.292(2)	not given	
$[Fe^0(CO)_4\{(OPr)_2Si(hmpa)\}]$ (colourless)	m $P2_1/n$ 4	10.192(2) 14.780(3) 17.095(3)	98.44(1)	FeC_4Si	OC_{eq} 1.757(7,10) OC_{ax} 1.778(7) Si_{ax} 2.261(2)	C_{eq},C_{eq} 119.2(3,2.1) C_{eq},C_{ax} 95.1(3,2.0) C_{eq},Si_{ax} 84.9(2,1.5) C_{ax},Si_{ax} 176.5(2)	53
$[Fe^0(CO)_4\{(SBu^t)_2Si(hmpa)\}]$ (colourless) (at 223 K)	or $P2_12_12_1$ 4	13.066(2) 12.155(1) 17.181(3)		FeC_4Si	OC_{eq} not given OC_{ax} not given Si_{ax} 2.278(1)	not given	54
$[Fe^0(CO)_4(AsCBu^t)_4]$ (yellow)	m $P2_1/m$ 2	10.281(4) 13.572(1) 10.783(1)	104.59(2)	FeC_4As	OC_{eq} 1.794(9,8) OC_{ax} 1.772(9) As_{ax} 2.370(1)	not given	55
$[Fe^0(CO)_4\{As(o\text{-}tolyl)_3\}]$ (white)	m $P2_1/n$ 4	10.265(2) 10.517(1) 21.697(3)	99.16(2)	FeC_4As	OC_{eq} 1.787(7,0) OC_{ax} 1.772(7) As_{ax} 2.393(1)	C_{eq},C_{eq} 120.0(3) C_{eq},C_{ax} 90.2(3) C_{eq},As_{ax} 89.8(2)	26
$[Fe^0(CO)_4SbPh_3]$ (yellow)	tr P-1 2	10.557(1) 10.043(1) 11.230(1)	114.22(1) 91.95(1) 92.25(1)	FeC_4Sb	OC_{eq} 1.787(4,5) OC_{ax} 1.765(4) Sb_{ax} 2.472(1)	C_{eq},C_{eq} 119.9(2,3.5) C_{eq},C_{ax} 91.7(2,1.4) C_{eq},Sb_{ax} 88.3(2,7) C_{ax},Sb_{ax} 178.1(4)	56

Table 2. Continued 8

(ppn)[$Fe^{-II}(CO)_4BiPh_2$] (yellow)	m Cc 4	13.08(1) 15.832(9) 23.18(1)	99.53(6)	FeC_4Bi	OC_{eq} 1.80(3,3) OC_{ax} 1.70(3) Bi_{ax} 2.676(1)	C_{eq},C_{eq} 119(1,7) C_{eq},C_{ax} 95(1,3) C_{eq},Bi_{ax} 85.0(9,5.8) C_{ax},Bi_{ax} 172(1)	57
[$Fe^0(CO)_2(Bu^tNC)(tb)$] (dark blue)	m C2/c 8	29.967(5) 6.273(1) 25.481(4)	110.37(2)	FeC_3N_2	OC 1.786(7,8) Bu^tNC 1.840(7) tb,N 1.932(5,1)	C,C 95.7(3) NC,C 90.2(3,3.0) N,N 81.0(2) C,N 98.2(3,5.6) 117.3(3) 147.0(3) NC,N 90.0(3) 163.1(3)	58
trans-[$Fe^0(CO)_3(PPh_2Me)_2$] (pale yellow)	m $P2_1/c$ 4	15.551(6) 9.7024(12) 17.320(6)	91.081(11)	FeC_3P_2	OC_{eq} 1.765(6,9) P_{ax} 2.2059(14,30)	C_{eq},C_{eq} 120.00(24,5.66) C_{eq},P_{ax} 90.09(17,2.94) P_{ax},P_{ax} 173.95(6)	59
trans-[$Fe^0(CO)_3\{P(OMe)_3\}_2$] (not given)	m C2/c 4	11.964(4) 11.208(4) 13.406(5)	108.80(4)	FeC_3P_2	OC_{eq} 1.762(7,2) P_{ax} 2.155(1)	C_{eq},C_{eq} 120.2(2,4) C_{eq},P_{ax} 89.7(2,5)	60
trans-[$Fe^0(CO)_3(PPh_3)_2$] (orange)	or Pbca 8	18.216(5) 17.038(1) 21.804(3)		FeC_3P_2	OC_{eq} 1.770(4,6) P_{ax} 2.217(1,3)	C_{eq},C_{eq} 119.99(17,3.57) C_{eq},P_{ax} 90.06(11,3.85) P_{ax},P_{ax} 172.56(4)	61
[$Fe^0(CO)_3(PPh_3)_2$] (pale yellow)	tr P-1 2	13.51(1) 15.34(1) 10.040(9)	93.85(8) 105.84(8) 63.92(6)	FeC_3P_2	OC_{eq} 1.772(8,17) P_{ax} 2.216(3,9)	C_{eq},C_{eq} 120.0(3,5.9) C_{eq},P_{ax} 90.1(2,4.7) P_{ax},P_{ax} 172.54(8)	62
[$Fe^I(CO)_3(PPh_3)_2$][PF_6]· $0.5CH_2Cl_2$ (deep green)	tr P-1 2	18.694(9) 11.434(5) 10.915(2)	63.40(3) 104.71(3) 109.49(4)	FeC_3P_2	OC_{eq} 1.81(1,1) P_{eq} 2.283(4,1) OC_{ax} 1.80(1)	C_{eq},C_{eq} 162.0(6) P_{eq},P_{eq} 163.4(1) C_{eq},P_{eq} 88.7(4,1.3) C_{eq},C_{ax} 99.0(6,2.4) P_{eq},C_{ax} 98.3(4,9)	63
[$Fe^0(CO)_3\{P(Me_2N)_3\}_2$] (colourless) (at 238 K)	m $P2_1/c$ 4	11.177(1) 15.778(1) 13.270(2)	90.09(1)	FeC_3P_2	OC_{eq} 1.759(3,2) P_{ax} 2.213(1,1)	C_{eq},C_{eq} 120.0(1,5.9) C_{eq},P_{ax} 90.1(1,5.3) P_{ax},P_{ax} 175.1(0)	37, 64

Table 2. Continued 9

$[Fe^0(CO)_3\{(CF_3)_2PNPPh_3\}_2]$ (not given)	m C2/c 4	15.5840(10) 17.418(2) 16.7730(10)	 95.470(0)	FeC_3P_2	OC_{eq} 1.782(3,0) P_{ax} 2.167(1,0)	C_{eq},P_{ax} 90.0(1,3.0) P_{ax},P_{ax} 177.8(1)	33
$[Fe^0(CO)_3(pmpe)_2]$ (not given)	m $P2_1/a$ 4	13.092(2) 15.853(2) 15.406(3)	 103.00(1)	FeC_3P_2	OC_{eq} 1.750(9,11) P_{ax} 2.208(2,3)	C_{eq},C_{eq} 120.0(4,13.3) C_{eq},P_{ax} 90.6(3,8.9) P_{ax},P_{ax} 162.3(1)	65
trans-$[Fe^0(CO)_3\{P(OCH_2)_3P\}_2]$ (not given)	or Pnma 4	11.516(5) 9.780(4) 17.065(6)		FeC_3P_2	OC_{eq} 1.80(2,2) P_{ax} 2.153(4,37)	C_{eq},C_{eq} 118.4(3) C_{eq},P_{ax} 90.5(4,4.7) P_{ax},P_{ax} 176.4(2)	66
trans-$[Fe^0(CO)_3(dpe)_2]$ (yellow)	m $P2_1/c$ 4	11.532(2) 20.454(2) 22.652(6)	 104.49(2)	FeC_3P_2	OC_{eq} 1.761(11,12) P_{ax} 2.209(3,0)	C_{eq},C_{eq} 120.7(5,2.4) C_{eq},P_{ax} 90.0(3,3.6) P_{ax},P_{ax} 175.64(12)	67
cis-$[Fe^0(CO)_3(dpe)_2]$ (orange yellow)	m $P2_1/c$ 4	10.064(1) 15.164(2) 17.379(2)	 101.17(1)	FeC_3P_2	OC_{eq} 1.759(6,9) P_{ax} 2.211(2,6)	C_{eq},C_{eq} 92.3(3,9) 117.7(3) C_{eq},P_{ax} 83.91(6) P_{ax},P_{ax} 121.1(2,5.6)	67
$[Fe^0(CO)_3(dppee)]$ (orange)	tr P-1 2	14.106(2) 10.683(1) 9.387(1)	101.730(11) 82.469(11) 94.505(10)	FeC_3P_2	OC_{eq} 1.755(7,2) P_{eq} 2.221(1,0) OC_{ax} 1.784(5)	C_{eq},C_{eq} 91.1(3) P_{eq},P_{eq} 74.4(1)[d] C_{eq},P_{eq} 90.5(2,2.7) C_{eq},C_{ax} 104.5(3,2) P_{eq},C_{ax} 103.5(2,1.0)	68
$[Fe^0(CO)_3(dppe)]$ (yellow)	m $P2_1/n$ 4	12.260(6) 15.890(7) 14.227(6)	 110.57(3)	FeC_3P_2	OC 1.768(5,10) P 2.224(2,2)	C,C 94.0(2,6.4) 112.6(2) P,P 84.1(1)[e] C,P 91.2(2,4.6) 104.0(2) 143.3(2) 164.1(2)	69
$[Fe^0(CO)_3(pp)]$ (dark green)	m $P2_1/c$ 4	15.120(5) 8.364(2) 21.812(8)	 93.00(3)	FeC_3P_2	OC_{eq} 1.767(4,23) P_{eq} 2.208(1) OC_{ax} 1.758(4) P_{ax} 2.219(1)	C_{eq},C_{eq} 115.0(2) C_{eq},P_{eq} 122.6(1,6.0) C_{eq},C_{ax} 92.0(2,3.8) C_{eq},P_{ax} 91.8(1,2.6) P_{eq},P_{ax} 87.9(1)[c] C_{ax},P_{ax} 169.8(1)	70

Table 2. Continued 10

$[Fe^{0}(CO)_3(hpfb)]$	m	9.473(4)		FeC_3P_2	OC_{eq}	1.747(14,33)	C_{eq},C_{eq}	113.2(7)	71a
(ruby-red)	Pc	11.026(4)	97.50(3)		P_{eq}	2.250(4)	C_{eq},P_{eq}	123.4(5,4.4)	
	2	15.561(9)			OC_{ax}	1.722(14)	C_{eq},C_{ax}	90.4(6,2.7)	
					P_{ax}	2.268(4)	P_{eq},C_{ax}	87.5(5)	
							P_{eq},P_{ax}	87.8(1)[e]	
							C_{eq},P_{ax}	92.6(5,4.7)	
							C_{ax},P_{ax}	169.7(5)	
$[Fe^{0}(CO)_3(dppt)]$	m	19.094(1)		FeC_3P_2	OC	1.657(14)	C,C	88.9(6,2.0)	71b
(yellow)	Cc	10.455(1)	114.89(3)			1.745(12,18)		128.4(5)	
	4	15.786(3)			P	2.253(3,3)	P,P	97.05(12)	
							C,P	88.7(4,2.1)	
								115.8(4,1.2)	
								171.8(4)	
$[Fe^{0}(CO)_3(dppea)]$	m	35.741(8)		FeC_3P_2	OC_{eq}	1.770(3,5)	C_{eq},C_{eq}	118.0(1)	72
(yellow)	C2/c	10.008(2)	95.72(1)		P_{eq}	2.193(1)	C_{eq},P_{eq}	120.3(1,6.9)	
	8	15.049(2)			OC_{ax}	1.767(3)	C_{eq},C_{ax}	94.4(1,2.1)	
					P_{ax}	2.200(1)	P_{eq},C_{ax}	93.7(1)	
							P_{eq},P_{ax}	70.8(0)[d]	
							C_{eq},P_{ax}	94.0(1,2.3)	
							C_{ax},P_{ax}	162.9(1)	
$[Fe^{0}(CO)_3(PcyPh_2)_2]$	m	12.393(5)		FeC_3P_2	OC_{eq}	1.771(5,12)	C_{eq},C_{eq}	120.7(2,2.6)	73
(light yellow)	$P2_1/n$	15.811(6)	108.00(2)		P_{ax}	2.218(1,0)	C_{eq},P_{ax}	89.7(1,2.8)	
(at 100 K)	4	18.029(7)					P_{ax},P_{ax}	178.45(6)	
$[Fe^{0}(CO)_3(atpd)]\cdot 0.5MeCN$	m	14.408(4)		FeC_3P_2	OC_{eq}	not given	P_{eq},P_{ax}	85.4(1)[e]	74
(red)	$P2_1/n$	26.949(6)	109.46(2)		P_{eq}	2.135(2)	C_{ax},P_{ax}	170.3	
	4	18.831(4)			OC_{ax}	not given			
					P_{ax}	2.253(2)			

Table 2. Continued 11

[$Fe^{0}(CO)_2(CS)(PPh_3)_2$][c] (yellow)	m $P2_1$ 4	18.821(5) 12.113(3) 18.149(5)	 117.76(6)	FeC_3P_2	OC_{eq} 1.756(9,10) SC_{eq} 1.550(7) P_{ax} 2.205(2,27)	C_{eq},C_{eq} 113.0(4) SC_{eq},C_{eq} 123.1(4,1.7) C_{eq},P_{ax} 90.8(2,2.3) SC_{eq},P_{ax} 88.6(2,2.0) P_{ax},P_{ax} 176.08(8)	75
				FeC_3P_2	OC_{eq} 1.755(8,23) SC_{eq} 1.563(8) P_{ax} 2.209(2,24)	C_{eq},C_{eq} 129.4(4) SC_{eq},C_{eq} 115.3(4,5.0) C_{eq},P_{ax} 89.0(2,6) SC_{eq},P_{ax} 92.5(2,2.8) P_{ax},P_{ax} 175.02(8)	
[$Fe^{0}(CO)_2(CS)\{P(OPh)_3\}_2$] (yellow)	tr P-1 2	11.040(6) 9.915(5) 17.247(10)	92.4(1) 98.0(1) 100.0(1)	FeC_3P_2	OC_{eq} 1.82(1) SC_{eq} 1.79(1,2) P_{ax} not given	C_{eq},C_{eq} 120.0(5,2.3) C_{eq},P_{ax} 90.0(3,4.3) P_{ax},P_{ax} 177.3(1)	76
[$Fe^{0}(CO)_3$(diars)] (pale yellow)	or Pnma 4	16.73(3) 10.37(3) 9.71(3)		FeC_3As_2	OC_{eq} 1.81 As_{eq} 2.31 OC_{ax} 1.68 As_{ax} 2.36	C_{eq},C_{eq} 122 C_{eq},C_{ax} 92 C_{eq},As_{eq} 119 C_{eq},As_{ax} 89 As_{eq},As_{ax} 84[e] C_{ax},As_{ax} 92	77
[$Fe^{0}(CO)_2\{P(CF_3)_3\}_3$] (orange)	hx $P6_3$ 2	10.625(10) 12.955(15)	90 90 120	FeP_3C_2	P_{eq} 2.151(4) C_{ax} 1.740(21)	C_{ax},P_{eq} 90.0(4)	78
[$Fe^{0}(CO)_2\{P(OMe)_3\}_3$] (yellow)	m C2/c 8	29.00(3) 9.269(4) 16.16(2)	 106.2(1)	FeP_3C_2	P_{eq} 2.153(3) OC_{eq} 1.75(1,0) P_{ax} 2.146(3,1)	C_{eq},C_{eq} 130.0(5) C_{eq},P_{eq} 115.0(4,6) C_{eq},P_{ax} 87.7(4,3.1) P_{ax},P_{ax} 168.3(1)	79
[$Fe^{0}(CO)_2\{P(OEt)_3\}_3$] (colourless)	m $P2_1/c$ 4	15.454(8) 13.100(6) 15.836(6)	 101.81(4)	FeP_3C_2	P_{eq} 2.187(2) OC_{eq} 1.668(8,4) P_{ax} 2.172(2,6)	C_{eq},C_{eq} 123.6(3) C_{eq},P_{eq} 118.2(3,2.8) C_{eq},P_{ax} 87.9(2,3.3) P_{ax},P_{ax} 94.9(1,1.0)	80

Table 2. Continued 12

[$Fe^0H(CO)_3$(eph)] (not given)	tr P-1 1	9.667 9.547 10.179	94.44 110.03 117.93	FeC_3HP	OC_{eq} P_{eq} OC_{ax} H_{ax}	1.737(5,6) 2.159(1) 1.749(5) 1.57(6)	C_{eq},C_{eq} C_{eq},P_{eq} C_{eq},H_{ax} C_{eq},C_{ax} P_{eq},H_{ax} C_{ax},H_{ax}	116.6(2) 96.8(-) 81.4(21,2.1) 97.1(2,3) 86.4(20) 176.8(21)	81
(NEt_4)[trans-$HFe^0(CO)_3(PPh_3)$] (yellow)	c Pa3 8	17.687(8)	90.0	FeC_3HP	OC_{eq} P_{ax} H_{ax}	1.719(9) 2.187(1) not given	C_{eq},C_{eq} C_{eq},P_{ax}	117.3(2) 99.5(3)	82
(ppn)[cis-$HFe^0(CO)_3${$P(OPh)_3$}] (yellow)	tr P-1 2	10.491(2) 13.963(4) 18.687(6)	108.19(2) 105.47(2) 93.16(2)	FeC_3HP	OC_{eq} P_{eq} OC_{ax} H_{ax}	1.732(6,5) 2.083(2) 1.728(10) 1.560(70)	C_{eq},C_{eq} C_{eq},P_{eq} C_{eq},H_{ax} P_{eq},H_{ax} P_{eq},C_{ax} C_{eq},C_{ax} C_{ax},H_{ax}	121.0(4) 115.0(3,2.4) 78.9(19,2.1) 82.0(22) 95.7(3) 102.2(4,9) 177.4(22)	83
(ppn)[$Fe^0(CO)_3(PPh_3)SiCl_3$] (not given)	m $P2_1/c$ 4	16.087(3) 15.334(2) 22.804(2)	109.45(1)	FeC_3PSi	OC_{eq} P_{ax} Si_{ax}	1.754(8,11) 2.207(3) 2.197(3)	C_{eq},C_{eq} C_{eq},P_{ax} C_{eq},Si_{ax} P_{ax},Si_{ax}	119.7(4,2.4) 93.2(3,3.9) 86.8(4,1.5) 174.19(9)	84
(ppn)[$Fe^0(CO)_3${$P(OPh)_3$}$SiCl_3$] (not given)	m $P2_1/c$ 4	18.623(3) 13.449(6) 23.447(6)	109.91(2)	FeC_3PSi	OC_{eq} P_{ax} Si_{ax}	1.73(1,3) 2.121(4) 2.206(4)	C_{eq},C_{eq} C_{eq},P_{ax} C_{eq},Si_{ax} P_{ax},Si_{ax}	119.7(8,6.0) 93.4(5,4.8) 86.8(5,1.6) 173.7(1)	84
[$Fe^0(CO)_3${$P(OEt)(NMe_2)_2$}- {$Si(OEt)_2NHMe_2$}] (pale brown) (at 183 K)	m $P2_1/a$ 8	15.302(8) 17.129(3) 18.42(2)	109.38(4)	FeC_3PSi	OC_{eq} P_{ax} Si_{ax}	1.757(8,10) 2.186(2) 2.218(2)	C_{eq},C_{eq} P_{ax},Si_{ax}	119.9(3,2.1) 173.4(1)	85
(NEt_4)[Fe^{II}(CO)(PS3*)] (not given)	c $P2_13$ 4	15.8730(8)		FeS_3CP	S_{eq} OC_{ax} P_{ax}	2.290(4,0) 1.88(3) 2.165(7)	S_{eq},S_{eq} S_{eq},C_{ax} S_{eq},P_{ax} $C_{ax}...P_{ax}$	119.35(3) 94.6(1) 85.4(1)[e] 180.00	86

Table 2. Continued 13

Compound	Crystal class Space group Z	a b c	α β γ	Chromophore	Fe–L		L–Fe–L		Ref.
$[Fe^{0}(CO)(NO)(MeCN)(PPh_3)_2]$. $BF_4 \cdot CH_2Cl_2$ (red orange)	tr P-1 2	9.7782(7) 10.2832(7) 20.5752(14)	92.903(6) 99.535(6) 91.930(6)	FeN_2P_2C	CN_{eq} N/C_{eq} P_{ax}	1.989(3) 1.706(4,17) 2.277(1,5)	N/C, N/C CN,N/C P,P	117.6(2) 121.3(2,2.5) 170.20(4)	87
$[Fe^{0}(CO)_2(PPh_3)\{S_2C_2(SMe)_2\}]$ (blue)	m $P2_1/a$ 4	24.144(6) 10.017(2) 10.745(2)	 97.38(2)	FeC_2S_2P	OC_{eq} S_{eq} P_{eq} OC_{ax}	1.788(1) 2.180(3,17) 2.244(3) 1.776(13)	S_{eq},S_{eq} C_{eq},S_{eq} C_{eq},P_{eq} S_{eq},P_{eq} S_{eq},C_{ax} P_{eq},C_{ax} C_{eq},C_{ax}	87.8(1)[e] 87.4(3) 150.8(4) 90.0(3) 89.1(1) 168.6(1) 104.0(4,5.1) 92.5(3) 100.0(5)	88
$[Fe^{0}(CO)_2\{P(NMe_2)_3\}$- $SC\{P(NMe_2)_3\}C(O)S]$ (violet)	m $P2_1/c$ 4	13.124(1) 14.131(2) 14.070(1)	 91.71(1)	FeC_2S_2P	OC_{eq} S_{eq} P_{eq} OC_{ax} S_{ax}	1.790(9) 2.150(2) 2.214(3) 1.78(1) 2.240(2)	C_{eq},S_{eq} C_{eq},P_{eq} S_{eq},P_{eq} C_{eq},C_{ax} S_{eq},C_{ax} P_{eq},C_{ax} C_{eq},S_{ax} S_{eq},S_{ax} P_{eq},S_{ax} C_{ax},S_{ax}	140.5(3) 101.4(3) 118.1(1) 92.2(4) 85.8(3) 94.1(3) 89.0(3) 86.9(1)[c] 95.4(1) 169.9(3)	89a
$[Fe(CO)_2\{(MeS)_2C_2S_2\}(PPh_3)]$	m $P2_1/a$ 4	24.144(6) 10.017(2) 10.745(2)	 97.38(2)	FeC_2S_2P	OC S Ph_3P	1.781(13,7) 2.179(3,17) 2.244(3)	C,C S,S C,S C,P S,P	100.0(5) 87.8(1)[5] 95.8(4,10.6) 150.8(4) 91.2(3,1.3) 87.8(1) 168.6(1)	89b
$[Fe^{0}(CO)_2(NO)(PPh_3)_2]$. $BF_4 \cdot CH_2Cl_2$ (light brown)	m $P2_1/n$ 4	17.247(3) 13.754(2) 16.914(2)	 104.26(1)	FeC_2P_2N	C/N_{eq} P_{ax}	1.762(4,7) 2.280(1,2)	$C/N_{eq},C/N_{eq}$ $C/N_{eq},P_{ax}$ P_{ax},P_{ax}	120.0(2,1.9) 90.0(1,1.7) 177.45(5)	90
$[Fe^{0}(CO)_2(PEt_3)_2(EtCN)]$ (yellow) (at 238 K)	m $P2_1/c$ 4	12.236(4) 7.921(3) 20.176(8)	 106.56(3)	FeC_2P_2N		not given		not given	91

Table 2. Continued 14

Compound	Cryst. cl. / Space gr. / Z	a, b, c (Å)	α, β, γ (°)	Chromophore	Fe–L	(Å)	L–Fe–L	(°)	Ref.
$[Fe^{0}(CO)_2(PPh_3)_2(N_2Ph)]BF_4$ (red orange)	m $P2_1/c$ 4	13.447(4) 14.260(4) 22.755(16)	 113.29(2)	FeC_2P_2N	OC_{eq} N_{eq} P_{ax}	1.793(9,26) 1.702(6) 2.264(2,3)	C_{eq},C_{eq} C_{eq},N_{eq} C_{eq},P_{ax} N_{eq},P_{ax} P_{ax},P_{ax}	107.6(3) 126.2(3,4.8) 90.3(2,1.9) 89.67(18,48) 175.85(8)	92
$[Fe^{0}(CO)_2\{P(OPh)_3\}_2(SO_2)]$ (orange)	tr P1 2	11.027(5) 11.374(5) 15.238(8)	88.3(1) 97.7(1) 108.6(1)	FeC_2P_2S	OC_{eq} O_2S_{eq} P_{ax}	1.781(15,11) 2.051(4) 2.172(5,8)	C_{eq},C_{eq} C_{eq},S_{eq} C_{eq},P_{ax} S_{eq},P_{ax} P_{ax},P_{ax}	116.9(7) 121.6(5,4) 89.0(5,3) 91.9(2,2) 176.2(2)	93a
$[Fe^{0}(CO)_2\{P(o\text{-}tolO)_3\}_2(SO_2)]$ (not given)	m C2/c 4	19.873(11) 10.725(5) 20.798(11)	 100.7(1)	FeC_2P_2S	P_{eq} O_2S_{eq} OC_{ax}	2.170(4,0) 2.049(6) 1.80(2,0)	P_{eq},P_{eq} P_{eq},S_{eq} P_{eq},C_{ax} S_{eq},C_{ax} C_{ax},C_{ax}	123.3(2) 118.3(1) 90.2(5,2) 90.8(5) 178.4(7)	93a
$[Fe^{0}(CO)_2\{P(OC_6H_4Me\text{-}o)_3\}_2\cdot$ $(SO_2)]$ (orange)	m C2/c 4	19.873(11) 10.725(5) 20.798(11)	 100.7(1)	FeC_2P_2S	P_{eq} O_2S_{eq} OC_{ax}	2.167(4,0) 2.055(7) 1.811(18,0)	P_{eq},P_{eq} P_{eq},S_{eq} P_{eq},C_{ax} S_{eq},C_{ax} C_{ax},C_{ax}	123.3(1) 118.4(1) 89.5(6,3) 91.0(5) 178.0(8)	93a
$[Fe^{0}(CO)_2(\eta^2\text{-}CS_2)(PMe_3)(PPh_3)]$ (yellow orange)	m Pc 2	9.309(4) 13.640(12) 11.390(5)	 120.43(5)	FeC_2P_2S	OC_{eq} CS_{eq} SC_{eq} P_{ax}	1.791(7,20) 2.334(2) 1.983(8) 2.266(2,14)	C_{eq},C_{eq} C_{eq},SC_{eq} C_{eq},CS_{eq} SC_{eq},CS_{eq} C_{eq},P_{ax} SC_{eq},P_{ax} CS_{eq},P_{ax} P_{ax},P_{ax}	109.7(3) 92.3(3) 157.2(2) 125.1(1,11.8) 44.8(2) 89.7(2,4.6) 91.3(2,2.5) 90.2(0,1.3) 176.8(0)	93b
$[Fe^{0}(CO)_2\{P(OPh)_3\}_2(SO_2)]$ (red brown) (at 213 K)	tr P-1 2	11.339(2) 15.220(2) 10.989(2)	97.45(2) 108.25(2) 88.30(2)	FeC_2P_2S	OC_{eq} O_2S_{eq} P_{ax}	1.782(3,1) 2.059(1) 2.173(1,1)	C_{eq},C_{eq} C_{eq},S_{eq} C_{eq},P_{ax} S_{eq},P_{ax} P_{ax},P_{ax}	114.7(1) 122.6(1,2) 89.1(1,5) 91.7(0,2) 176.6(0)	94

Table 2. Continued 15

$[Fe^{II}(CO)\{P(OMe)_3\}_2(anpedt)]$	m	8.176(1)		FeP_2S_2C	P_{eq}	2.140(1,15)	P_{eq},P_{eq}	94.1(0)	95
(violet red)	$P2_1/n$	29.308(3)	99.06(1)		S_{eq}	2.155(1)	P_{eq},S_{eq}	132.9(0,8.0)	
	4	11.021(1)			S_{ax}	2.196(1)	P_{eq},S_{ax}	92.9(0,6)	
					OC_{ax}	1.756(5)	P_{eq},C_{ax}	91.5(1,8)	
							S_{eq},S_{ax}	88.2(0)[e]	
							S_{eq},C_{ax}	85.4(1)	
							S_{ax},C_{ax}	173.6(1)	
$[Fe^0H(CO)(NO)(PPh_3)_2]$[c]	or	13.443(1)		FeP_2HNC	OC_{eq}	1.760(6)	C_{eq},N_{eq}	120.6(3)	96
(orange)	$Pbc2_1$	23.101(2)			ON_{eq}	1.666(5)	C_{eq},P_{ax}	94.5(2,9)	
(at 115 K)	8	20.284(4)			P_{ax}	2.188(2,0)	N_{eq},P_{ax}	103.9(2,7)	
					H_{eq}	not located	P_{ax},P_{ax}	140.9(1)	
				FeP_2HNC	OC_{eq}	1.745(6)	C_{eq},N_{eq}	125.8(3)	
					ON_{eq}	1.680(6)	C_{eq},P_{ax}	94.7(2,1.9)	
					P_{ax}	2.177(2,3)	N_{eq},P_{ax}	102.5(2,4.1)	
					H_{eq}	not located	P_{ax},P_{ax}	140.9(1)	
$[Fe^0H(CO)(NO)(PMe_3)_2]$	or	9.437(2)		FeP_2HNC	H_{eq}	1.497(24)	H_{eq},N_{eq}	117.5(3)	97
(not given)	Pnma	11.457(3)			OC_{eq}	1.807(11)	H_{eq},C_{eq}	120.8(3)	
	4	12.438(2)			ON_{eq}	1.619(9)	N_{eq},C_{eq}	121.7(5)	
					P_{ax}	2.184(1,4)	H_{eq},P_{ax}	77.7(9,4.2)	
							N_{eq},P_{ax}	95.7(3,1)	
							C_{eq},P_{ax}	96.3(2,1.6)	
							P_{ax},P_{ax}	155.4(1)	
$[Fe^0Cl(CO)(NO)(PMe_3)_2]$	or	7.416(2)		FeP_2NCCl	OC_{eq}	1.727(5)	N_{eq},Cl_{eq}	116.4(1)	97
(not given)	$P2_12_12_1$	12.296(4)			ON_{eq}	1.682(4)	C_{eq},Cl_{eq}	131.6(2)	
	4	21.282(6)			Cl_{eq}	2.349(2)	N_{eq},P_{ax}	98.5(1,1)	
					P_{ax}	2.257(2,10)	C_{eq},P_{ax}	87.7(2,2)	
							Cl_{eq},P_{ax}	85.4(1,4)	
							P_{ax},P_{ax}	163.0(1)	

Table 2. Continued 16

$[FeH_3(CO)(dppe)Si(OEt)_3]$[f] (colourless)	FeP_2HCSi	P_{eq}	2.209(4,12)	P_{eq},P_{eq}	86.6(1)[e]	98
		Si_{eq}	2.258(4)	P_{eq},Si_{eq}	128.2(2,3.6)	
		OC_{ax}	1.76(1)	Si_{eq},C_{ax}	106.1(5)	
		H	not given	P_{eq},C_{ax}	101.7(5,5.8)	
	FeP_2HCSi	P_{eq}	2.199(4,2)	P_{eq},P_{eq}	87.5(1)[e]	
		Si_{eq}	2.246(4)	P_{eq},Si_{eq}	127.5(2,7)	
		OC_{ax}	1.76(1)	Si_{eq},C_{ax}	112.6(5)	
		H	not given	P_{eq},C_{ax}	97.2(5,1.3)	
	FeP_2HCSi	P_{eq}	2.207(4,10)	P_{eq},P_{eq}	87.2(1)[e]	
		Si_{eq}	2.246(4)	P_{eq},Si_{eq}	127.5(2,1.9)	
		OC_{ax}	1.75(1)	Si_{eq},C_{ax}	112.3(5)	
		H	not given	P_{eq},C_{ax}	97.6(5,7)	

a Where more than one chemically equivalent distance or angle are present the mean values is tabulated. The first number in parenthesis is e.s.d.'s and the second in maximum deviation from the mean value.
b The chemical indentity of coordinated atom/ligand is specified in these columns.
c There are two crystallographically independent molecules.
d Four - membered metallocyclic ring.
e Five - membered metallocyclic ring.
f There are three crystallographically independent molecules.

Table 3. Crystallographic and Structural Data for Six-coordinated Iron Carbonyl Compounds[a]

COMPOUND (COLOUR)	CRYST. CL. SPACE GR. Z	a[Å] b[Å] c[Å]	α[°] β[°] γ[°]	CHROMO-PHORE	Fe–L [Å]	L–Fe–L [°]	REF.
[Fe(tpp)(py)(CO)] (purple)	m $P2_1/c$ 4	13.246(17) 19.555(26) 19.822(25)	 105.49(3)	FeN_5C	N_{eq}[b] 2.02(1,3) pyN_{ax} 2.10(1) OC[b] 1.77(2)	N,N[b] 90.0(5,1.3)[c] 89.5(6,1.0) 178.9(6,6) N,C 90.5(9,1.5) 177.5(8)	100
[Fe([14]annulene)(py)(CO)] (purple)	or Pbca 8	30..907(13) 9.619(5) 16.615(7)		FeN_5C	N_{eq} 1.941(2,3) pyN_{ax} 2.088(3) OC_{eq} 1.730(3)	N,N 83.8(1,6)[d] 88.6(1,1.3) 96.2(1,3.5)[e] 177.1(1,5) N,C 91.5(1,2.0) 177.2(1)	20
[Fe([14]annulene)(hz)(CO)] (purple)	m $P2_1/c$ 4	9.422(2) 11.015(1) 21.264(4)	 106.58(1)	FeN_5C	N_{eq} 1.943(4,11) hzN_{ax} 2.122(5) OC_{ax} 1.751(5)	N,N 83.2(2,6)d 86.8(2.1.5) 96.5(2,6)[e] 173.5(2,3) N,C 93.3(2,1.3) 179.2(2)	20
[Fe(C_3-cap)(CO)(1-Meim)]. .$CHCl_3$.2 H_2O (not given)	m $P2_1/c$ 4	18.422(12) 15.932(12) 22.54(2)	 90.76(2)	FeN_5C	N_{eq} 11.99(2,0) imN_{ax} 2.046(10) OC_{ax} 1.800(13)	N,N 90.0(4,9)[e] 177.6(4,4) N,C 88.8(4,1.9)	101a
[Fe(C_2-cap)(1-Meim)(CO)][f] (not given) at 123 K	tr P-1 4	18.022(2) 20.017(1) 20.691(2)	70.507(7) 76.232(1) 82.549(7)	FeN_5C	porN 1.991(6,11) MeimN 2.043(6) OC 1.742(7)	N,N 89.7(2,2.6) 90.0(2,7)[e] 178.6(3,6) N,C 90.4(3,5.2) 174.7(3)	101b
					porN 1.988(6,13) MeHimN 2.041(5) OC 1.748(7)	N,N 89.7(2,2.6) 90.0(2,1)[e] 178.5(5,1.4) N,C 90.7(3,3.1) 177.8(3)	

Table 3. Continued 2

[Fe(ddpp)(Meim)(CO)] (purple)	m C2/c 4	18.527(6) 19.603(7) 17.403(6)	 90.09(2)	FeN_5C	ddppN_{eq} 1.999(3) HimN_{ax} 2.062(5) OC_{ax} 1.728(6)	not given	102
[Fe(tpppiv8)(1-MeHim)(CO)]. .H_2O (dark blue)	tr P-1 2	13.110(3) 21.504(6) 12.009(3)	99.16(2) 111.71(2) 81.21(2)	FeN_5C	N_{ex} 1.991(4) N_{ax} 2.039(3) OC_{ax} 1.752(4)	not given	103
[Fe(tpppiv6)(1-MeHim)(CO)]. (1-MeHim) (dark blue)	m $P2_1/c$ 4	23.394(6) 11.878(3) 23.895(6)	 108.84(2)	FeN_5C	N_{ex} 1.981(3) N_{ax} 2.045(2) OC_{ax} 1.733(4)	not given	103
[Fe(tim)(acn)(CO)]$(PF_6)_2$ (not given)	m $P2_1/c$ 4	8.512(5) 18.047(7) 16.841(13)	 88.92(3)	FeN_5C	timN_{eq} 1.954(1) acnN_{ax} 1.976(3) OC_{ax} 1.779(3)	not given	104
[Fe(suboct[14]annulene)- (1-MeHim)(CO)] (not given)	m $P2_1/c$ 2	12.0(-) 8.40(-) 22.8(-)	 100	FeN_5C	OC not given N not given	not given	105
[Fe(OC_3Otpp)(1-Meim)- (CO)].1.5 tol (not given)	m $P2_1/n$ 4	16.221(7) 22.763(10) 17.133(6)	 92.70(3)	FeN_5C	tppN_{eq} 1.996(5,14) imN_{ax} 2.027(5) OC_{ax} 1.748(7)	N,N 90.0(2,8)[c] 90.2(2,1.1) 178.6(2,2) N,C 89.9(3,6.5) 174.5(3)	106
[Fe(OC_3Otpp)(1,2-Me_2Him)- (CO)].2 $CHCl_3$ (not given)	m C2/c 4	12.983(2) 33.705(6) 15.387(4)	 105.27(2)	FeN_5C	tppN_{eq} 2.00(4,15) imN_{ax} 2.102(6) OC_{ax} 1.713(8)	N,N 90.0(2,1)[c] 90.9(3,6.8) 178.1(2,1.8) N,C 89.07(12,91) 173.0(2)	106
[Fe$(dmgBBN)_2$(py)(CO)] (yellow)	m $P2_1/n$ 4	11.611(2) 19.512(3) 14.907(2)	 108.53(2)	FeN_5C	N_{eq} 1.891(4) pyN_{ax} 2.068(5) OC_{ax} 1.771(7)	N,C 175.0(6)	107
[Fe$(dmgBPh_2)_2$(py)(CO)]. CH_2Cl_2 (yellow)	or Pbcn 4	14.943(3) 15.890(3) 16.211(3)		FeN_5C	N_{eq} 1.886(5) pyN_{ax} 2.067(8) OC_{ax} 1.786(12)	N,C 180.0(1)	107

Table 3. Continued 3

$[Fe(dmgBF_2)_2(py)(CO)]\cdot$ CH_2Cl_2 (yellow)	or $P2_12_12_1$ 4	12.436(3) 12.246(2) 14.158(3)		FeN_5C	N_{eq} 1.892(4) pyN_{ax} 2.055(5) OC_{ax} 1.772(6)	N,C 179.0(6)	107
$[Fe(Et_2S_5)(CO)](BF_4)_2$ (orange)	tr P-1 2	10.475(4) 11.255(4) 13.086(5)	71.98(3) 66.48(4) 88.13(3)	FeS_5C	S 2.246(4,22) OC 1.777(12)	S,S 88.5(1,8)[d] S,C 90.0(3) 175.9(3)	108
[Fe(dpttd)(CO)] (red)	m $P2_1/c$ 4	7.406(1) 12.815(1) 19.329(2)	 98.39(1)	FeS_5C	S 2.282(1,29) CO 1.764(3)	S,S, 89.38(3,2.72)[d] 176.68(3,1.29) S,C 90.4(1,6)	109
$[Fe(4,5\text{-}\mu\text{-}B_6H_{10})(CO)_4]$ (colorless)	or Pbca 8	11.441(3) 11.696(4) 16.396(3)		FeC_4B_2	OC 1.806(3,13) B 2.180(4,3)	C,C 91.6(1,2.2) 110.2(1) 173.9(1) C,B 89.7(1,4.0) 102.9(1) 149.0(1,2.4) B,B 48.4(1)[e]	110
$(NBu^n_4)[Fe(B_7N_{12})(CO)_4]$ (deep amber)	m $P2_1/c$ 4	10.840(5) 11.629(6) 23.698(10)	 90.15(3)	FeC_4B_2	OC 1.775(12,44) B 2.207(8,11)	C,C 91.6(5,2.4) 112.3(4) 173.7(4) C,B 87.3(4,4.0) 100.2(4,1.7) 147.2(4,2.0) B,B 47.4(3)[e]	111
$[Fe\{B(4\text{-}Bu^tcat)\}_2(CO)_4]$ (not given)	or Pnma 4	11.991(7) 32.509(7) 6.352(9)		FeC_4B_2	OC 1.807(9,16) B 2.028(7)	C,C 96.0(4,3.0) 166.0(4) C,B 86.2(3,2.9) B,B 82.8(4)	112

Table 3. Continued 4

[Fe(η^3-S_2CPcy_3)$(CO)_3$][f] (deep red)	m $P2_1/n$ 8	22.409(9) 10.749(7) 22.132(7)	 112.36(4)	FeC_4S_2	OC 1.79(1,5) C 1.96(1) S 2.311(3,11)	C,C 96.7(6,4.7) 111.5(5,2) 135.1(5) C,S 47.4(3,6)[e] 90.9(4,2.3) 102.5(4,1.3) 157.4(4,6) S,S 78.6(1)[g]	113
				FeC_4S_2	OC 1.80(1,3) C 1.99(1) S 2.320(3,9)	C,C 97.2(5,3.8) 110.6(5,4) 135.8(5) C,S 47.6(3,3)[e] 90.2(4,1.8) 102.6(4,1) 157.1(4,2) S,S 78.4(1)[g]	
[Fe(dppe)(CNp-tol)(p-tolN=C-Np-tol-C=Np-tol)(CO)] (not given)	tr P-1 2	14.815(7) 15.428(8) 12.797(7)	107.92(3) 98.27(2) 94.42(3)	FeC_4P_2	OC 1.756(11) tolNC 1.829(9) $(tol)_3C$ 1.999(9,3) dppe 2.247(4,6)	not given	114
cis-[Fe$(SiMe_3)_2$$(CO)_4$] (pale yellow)	or Pnma 4	13.358(14) 17.591(18) 6.635(12)		FeC_4Si_2	OC 1.775(7,18) Si 2.456(2,0)	C,C 89.5(4) 103.7(3,3) 141.2(1) C,Si 79.3(2,6) Si,Si 111.8(1)	115
[Fe$(SiPh_2CEt)_2$$(CO)_4$] (yellow orange)	tr P-1 2	11.748(2) 17.246(3) 8.602(3)	86.70(2) 110.94(2) 107.68(2)	FeC_4Si_2	OC 1.781(9,24) Si 2.4113(26, 64)	C,C 100.9(4) 167.5(4) C,Si 85.3(3,1.9) 170.5(4,1.2) Si,Si 81.95(2)	116
[Fe(η^2-$\{(SiMe_3)_2CH\}_2Sb_2$)-$(CO)_4$] (orange)	m $P2_1/n$ 4	19.958(4) 6.635(1) 24.408(6)	 103.51(2)	FeC_4Sb_2	OC 1.801(8,21) Sb 2.715(1,4)	C,C 95.4(4,5.7) 167.1(3) C,Sb 89.3(3,8.5) 160.2(3,9) Sb,Sb 61.46(2)[e]	117

Table 3. Continued 5

$[Fe(BiPh_2)_2(CO)_4]$ (orange-yellow)	m $P2_1/n$ 4	12.009(3) 13.074(2) 17.264(2)	96.78(1)	FeC_4Bi_2	OC 2.23(1,3) Bi 2.827(2,5)	C,C 96.9(7,1.9) 158.2(7) Bi,Bi 90.39(6) C,Bi 84.0(5,3.1) 177.6(5,1.2)	57
$(NMe_3CH_2Ph)_2[FeBi_2Cl_6$- $(CO)_4]$ (orange)	or Pba2 2	14.562(3) 17.062(3) 7.2210(10)		FeC_4Bi_2	OC 1.84(2,1) Bi 2.684(13,11)	C,C not given Bi,Bi 83.4(5,5)	118
$[Fe(S_4)(CO)_2]$ (dark brown)	m $P2_1/c$ 4	12.933(6) 12.582(6) 11.519(4)	111.88(4)	FeS_4C_2	S 2.284(3,13) OC 1.793(9,8)	S,S 88.9(1,1.1)[d] 176.2(1) S,C 90.9(4,1.2) 178.6(4,3) C,C not given	119
$[Fe(Me_2S_4)(CO)_2](BF_4)_2$ (yellow)	m $P2_1/c$ 4	13.032(9) 11.231(15) 16.510(10)	97.98(6)	FeS_4C_2	S_{eq} 2.275(10,9) OC_{ax} 1.797(35,21)	S,S 86.2(3)[d] S,C 176.8(8) C,C 92.8(12)	108
$[Fe(Et\text{-}S_4)(CO)_2](BF_4)$. 1.5 CH_2Cl_2 (not given)	m $P2_1/n$ 4	11.430(5) 21.106(8) 12.082(5)	117.42(4)	FeS_4C_2	S_{eq} 2.289(2,12) OC_{ax} 1.800(7,3)	S,S 89.3(1)[d] S,C 174.3(2) C,C 93.4(3)	108
$[Fe(Et_2S_4)(CO)_2](BF_4)_2$ (yellow)	m $P2_1/c$ 4	14.171(5) 11.291(8) 16.179(5)	95.42(3)	FeS_4C_2	S_{eq} 2.281(2,8) OC_{ax} 1.814(4,4)	S,S 88.6(1)[d] S,C 177.0(1) C,C 92.4(1)	108
$[Fe(CH_3SC_6H_4S)_2(CO)_2]$ (not given)	or Pbca 8	11.274(2) 23.069(2) 13.991(2)		FeS_4C_2	S 2.296(1,25) OC 1.793)3,10)	S,S 85.1(1,1.0) 89.4(1,5)[d] 172.8(1) S,C 91.9(1,6.7) 172.4(1,2.9) C,C 94.1(1)	120

Table 3. Continued 6

$[Li(thf)_3][Fe(CH_3SC_6H_4S)_2$.	m	16.97(1)		FeS_4C_2	S 2.306(7,51)	S,S 87.1(2,1.3)[d]	120
(PhCO)(CO)]. thf	P2/c	17.17(2)	112.0(1)		PhOC 1.92(2)	90.7(2)	
(red)	8	27.42(1)			OC 1.71(2)	175.0(3)	
						S,C 89.5(5)	
						90.1(6)	
						176.8(5)	
						178.1(6)	
						C,C 92.9(9)	
$[Fe(dttd)(CO)_2]$	m	9.051(6)		FeS_4C_2	S 2.311(6,6)	S,S 88.76(-,95)[d]	121
(red)	C_2/c	11.66(1)	102.02(6)		OC 1.780(1)	173.45(-)	
	4	11.707(7)				S,C 90.86(-)	
						179.12(-)	
						C,C 88.58(-)	
$[Fe(Bu^t_4\text{-}dttd)(CO)_2]$. tol	m	17.22(1)		FeS_4C_2	S 2.261(3,16)	S,S 88.2(1)[d]	122a
(colorless)	C2/c	11.35(1)	100.59(7)		OC 1.764(7)	89.0(1,2.4)	
	4	19.76(2)				172.7(1)	
						S,C 91.5(2,2.3)	
						179.4(2)	
						C,C 90.3(5)	
$[Fe(CO_2MeS_4)(CO)_2]$	tr	7.841(8)	113.58(8)	FeS_4C_2	C 2.284(3,21)	S,S 89.4(1)[d]	122b
(red)	P-1	16.605(15)	98.17(8)		OC 1.759(8)	176.7(1)	
	4	18.920(2)	94.58(8)			S,C 89.8(3,1.2)	
						C,C 91.4(4)	
$[Fe(HBpz_3)(MeCO)(CO)_2]$	m	9.233(1)		FeN_3C_3	N 2.024(4,58)	N,N 86.8(1,1.9)[f]	123
(golden yellow)	$P2_1/c$	12.331(4)	109.64(1)		OC 1.765(6,7)	N,C 91.7(1,2.7)	
	4	15.507(2)			MeOC 1.968(5)	176.5(2,1.3)	
						C,C 89.1(2,1.2)	
fac-[K(18-crown-6)].	c	19.609(4)		FeC_3S_3	OC 1.802(7)	C,S 90.2(2,3.4)	124
$[Fe(SPh)_3(CO)_3]$	Pa3				PhS 2.350(2)	172.5(2)	
(not given)	8					S,S 85.7(7)	
fac-[Na(15-crown-5)].	tr	11.915(5)	65.77(5)	FeC_3S_3	OC 1.788(12,19)	C,C 96.1(5,1.1)	124
$[Fe(SPh)_3(CO)_3]$	P-1	12.391(3)	116.28(5)		PhS 2.341(3,4)	C,S 89.6(4,5.8)	
(not given)	2	14.337(3)	105.87(5)			169.6(4,4)	
						S,S 84.3(2,2)	

Table 3. Continued 7

fac-(ppn)[Fe(TePh)$_3$(CO)$_3$]	m	32.116(1)		FeC$_3$Te$_3$	OC 1.770(12,8)	C,C 96.2(6,1.1)	125
(dark purple)	C2/c	11.760(4)	94.942(3)		PhTe 2.6303(20,	C,Te 89.7(4,5.3)	
	8	29.027(10)			62)	169.9(4,2.1)	
						Te,Te 83.84(6,1.84)	
[Fe(pc)(dmf)(CO)]. dmf	m	9.86(1)		FeN$_4$OC	pcN 1.89(2,6)	N,N 90.0(9,8)[c]	126a
(purple)	P2$_1$/n	17.35(3)	87.9(2)		dmfO$_{ax}$ 2.07(1)	177.6(9,1)	
	4	19.79(4)			OC$_{ax}$ 1.72(2)	N,O 88.8(8,2.2)	
						N,C 91.2(11,2.9)	
						O,C 176.4(11)	
[Fe(deut)(thf)(CO)]	tr	11.524(3)	91.51(2)	FeN$_4$OC	deutN 1.955(5,2)	N,N 90.0(2,4)	126b
(purple)	P-1	14.915(5)	108.99(2)		2.005(4,2)	O,C 177.4(9)	
	2	10.583(2)	102.86(3)		thfO 2.127(4)		
					OC 1.706(5)		
[Fe(bmptee)(CO)][f]	m	22.608(5)		FeS$_4$OC	S 2.269(2,26)	S,S 89.2(1,3)[d]	127
(deep violet)	P2$_1$/c	20.256(4)	93.25(1)		O 2.076(5)	91.0(1)	
	8	7.947(1)			OC 1.721(9)	174.3(1,5.0)	
						S,O 84.7(2,1)[d]	
						91.2(2)	
						C,O 177.3(3)	
				FeS$_4$OC	S 2.261(2,34)	S,S 89.3(1,5)[d]	
					O 2.075(6)	90.1(1)	
					OC 1.706(10)	174.7(1,4.3)	
						S,O 85.2(2,2)[d]	
						90.3(2)	
						C,O 176.5(3)	

Table 3. Continued 8

[Fe(bmptea)(CO)][f] (dark red)	m $P2_1/c$ 8	22.447(6) 20.224(5) 8.052(2)	 92.45(2)	FeS_4NC	S 2.270(3,45) N 2.072(8) OC 1.753(12)	S,S 88.9(1,2)[d] 88.9(1) 175.0(1,2.6) S,N 86.4(2,1.0)[d] 90.1(2,1.4) 176.4(5) N,C 177.9(4)	127
				FeS_4NC	S 2.276(3,38) N 2.076(7) OC 1.744(11)	S,S 88.9(1,5)[d] 89.6(1) 174.5(1,3.4) S,N 85.7(2,5)[d] 90.6(2,1.3) N,C 177.9(4)	
[Fe(mftdm)(CO)] (white)	m $P2_1/c$ 4	7.539(4) 12.837(8) 19.319(10)	 95.86(4)	FeS_4NC	S 2.266(2,21) N 2.287(4) OC 1.732(4)	S,S 89.8(1)[d] 176.7(1) S,N 91.7(1)[d] S,C 90.2(2) N,C 176.8(2)	128
[Fe(dmmpte)(CO)I]FeI_4. CH_2Cl_2 (yellow)	tr P-1 4	8.562(2) 17.660(5) 22.030(7)	93.19(2) 94.46(2) 93.62(2)	FeS_4CI	S 2.268(5,24) OC 1.787(16) I 2.637(3)	S,S 88.7(5,0)[d] S,C 174.9(5) S,I 87.5(1) 176.3(1) C,I 88.5(5)	108
trans-[Fe{P(OEt)$_3$}$_4$(CO)H]. (BPh_4) (colorless)	or $P2_12_12_1$ 4	23.658(14) 23.671(13) 10.507(6)		FeP_4HC	P 2.186(4,17) H 1.64(16) OC 1.75(1)	P,P 90.3(2,9) 155.8(2) 176.9(2) P,H 79(5,3) 101(6) P,C 95.3(5,8.4) H,C 170(6)	129

Table 3. Continued 9

Compound (colour)	Crystal class / Space group / Z	a, b, c	α, β, γ	Chromophore	Distances	Angles	Ref.
[Fe(η^4-ehtp)(CO)Cl] (yellow)	m C2/c 8	36.975(6) 10.471(2) 26.159(5)	 134.12(3)	FeP_4CCl	P 2.2461(9,331. OC 798(7) Cl 2.362(2)	P,P 83.82(3)[d] 73.31(4)[g] 175.47(4) P,C 163.9(4) P,Cl 90.03(5,4.58) 164.83(5) C,Cl 99.1(4)	130
[Fe(enbt)(py)(CO)] (red black)	or $P2_12_12_1$ 4	13.323(8) 13.328(8) 11.747(7)		FeN_3S_2C	pyN 2.11(2) N 1.97(2,1) S 2.274(7,1) OC 1.74(3)	N,N 86.4(7)[d] 88.9(7,1.2) N,S 93.7(6,0)[c] 179.4(6,3) S,S 86.2(3) N,C 91.9(9,1) 178(1) S,C 88.7(8,4)	131
[Fe(AsPh$_3$)(CO)$_3$(H)$_2$] (colorless)	tr P-1 2	9.372(6) 10.465(9) 10.652(9)	108.12(3) 84.27(6) 87.16(6)	FeC_3H_2As	OC 1.787(8,18) H 1.35(-,0) As 2.304(1)	C,C 100.1(3,1.4) C,As 98.1(3,1.2) 151.6(3) H,As 81(-,2)	132
[Fe(η^2-OCH$_2$){P(OMe)$_3$}$_2$·(CO)$_2$] (not given)	m C2/c 8	32.66(7) 7.05(1) 14.84(3)	 101.6(2)	FeC_3P_2O	OC 1.76(2,3) OH$_2$C 2.03(2) ηO 2.00(1) P 2.211(4,9)	C,C 103.2(7) 108.5(6) 148.3(6) C,P 89.6(5,1.8) 90.9(4,1.8) C,O 38.3(5)[e] 110.1(6) 146.7(5) P,O 90.4(3,4)	133a

Table 3. Continued 10

[Fe($Si_2F_4CBu^tCH$)(PMe_3). $(CO)_3$][f] (not given)	or $Pca2_1$ 8	15.617(3) 8.161(3) 29.858(4)		FeC_3Si_2P	OC 1.735(22,13) Si 2.270(7,14) Me_3P 2.278(7)	C,C 95.3(10,1.1) Si,Si 81.64(24)[d] C,Si 86.7(7,4.5) 171.2(8) C,P 93.2(7,2.6) Si,P 92.99(22) 174.62(21)	133b
				FeC_3Si_2P	OC 1.800(23,27) Si 2.304(7,0) Me_3P 2.237(7)	C,C 94.4(10,1) 169.6(10) Si,Si 81.01(24)[d] C,Si 86.6(7,1.7) 169.2(7) C,P 93.8(7,3.2) Si,.P 94.11(24) 175.1(3)	
[Fe($Si_2F_4CBu^tCH$)(PPh_3). $(CO)_3$] (not given)	or $Pca2_1$ 4	17.490(8) 9.439(2) 17.332(5)		FeC_3Si_2P	OC 1.713(22,17) Si 2.286(6,27) Ph_3P 2.256(5)	C,C 96.2(10,2) 163.6(11) Si,Si 81.70(23)[d] C,Si 83.2(8,4.0) 160.9(7) C,P 94.6(7,4.2) Si,P 105.09(21) 172.05(21)	133b
[Fe($TePh_2$)$(CO)_3I_2$] (brown-red)	tr P-1 2	9.998(3) 10.139(4) 10.288(7)	103.32(4) 97.54(4) 112.46(3)	FeC_3I_2Te	OC 1.773(9,50) I 2.600(2,36) Te 2.585(2)	C,C not given C,I 86.7(3,1.3) 174.6(3) C,Te 95.1(3,1.3) 165.9(3) I,I 92.58(6) I,Te 83.55(7,23)	134a
(ppn)[Fe$(SePh)_2$(TePh)$(CO)_3$]. 0.5 thf (brown red)	m C2/c 8	32.035(8) 11.708(6) 28.909(6)	 95.00(2)	FeC_3Se_2Te	OC 1.750(11,13) Se/Te 2.509(2,20)	C,C 95.7(5,1.7) Se/Te 85.0(7,1.0) C,Se/ 89.6(4,5.9) Te 170.2(4,2.7)	134b

Table 3. Continued 11

(ppn)[Fe(TePh)$_2$(SePh)(CO)$_3$].	m	32.089(5)		FeC_3Se_2Te	OC 1.777(12,14)	C,C	95.9(5,1.2)	134b
0.5 thf	C2/c	11.745(2)	94.90(2)		Te/Se 2.568(2,26)	Te/Se	84.5(1,1.4)	
(brown red)	8	28.990(8)				C,Te/	89.6(3,5.3)	
						Se	170.2(4,2.0)	
cis-[Fe(Ph$_2$PCHCOPh)$_2$(CO)$_2$].	tr	13.090(4)	92.98(2)	$FeC_2O_2P_2$	OC 1.753(8,1)	C,C	92.9(2,2)	135a
thf. 0.5 C_6H_6	P-1	16.482(5)	108.87(2)		O 1.972(4,6)		177.8(6,1)	
(yellow)	2	10.274(3)	100.40(2)		P 2.255(2,9)	C,O	85.4(2)	
							175.6(2,1.4)	
						C,P	91.9(2,1.2)	
						O,O	88.8(1)	
						O,P	85.4(1,6)[d]	
						P,P	169.38(6)	
[Fe(dppa)$_2$(CO)$_2$]	m	36.293(2)		$FeO_2C_2P_2$	OC 1.770(1,14)	O,O	87.29(3)	135b
(yellow)	C2/c	7.5736(5)	104.35(1)		O 1.9813(7,33)	O,C	90.17(5,1.44)	
	8	21.082(1)			P 2.2571(3,47)		175.38(5,14)	
						O,P	82.1(1,6)[d]	
							88.6(1,1.0)	
						C,C	92.59(6)	
						C,P	94.42(4,1.55)	
						P,P	167.00(1)	
[Fe(qt)$_2$(CO)$_2$]	m	18.105(4)		$FeN_2C_2S_2$	OC 1.778(6,10)	N,N	98.1(2)	136
(brown)	C2/c	7.087(2)	104.03(1)		qtN 2.003(5,7)	C,C	92.3(3)	
	8	29.556(6)			qtS 2.333(3,15)	S,S	93.5(7)	
						N,C	96.2(2,9.5)	
							168.2(2)	
						N,S	70.5(1,1)	
							88.0(1)	
							161.2(1)	
						C,S	92.4(3,6.0)	
							176.2(2)	

Table 3. Continued 12

Compound	Cryst. cl. / Space gr. / Z	a, b, c	α, β, γ	Chromophore	Fe–L		L–Fe–L		Ref.
$[Fe(Sepy)_2(CO)_2]$[f]	tr	8.6691(10)	105.811(10)	$FeN_2C_2Se_2$	OC	1.772(9,15)	N,N	86.9(3)	137
(red-purple)	P-1	12.443(2)	107.533(8)		N	2.004(7,1)	N,C	91.5(3,1.3)	
	4	14.085(2)	92.075(10)		Se	2.450(2,8)		169.9(3,2)	
							C,C	91.9(4)	
							N,Se	72.1(2,2)[g]	
								97.2(2,1.4)	
							C,Se	95.0(3,3.9)	
							Se,Se	165.43(6)	
				$FeN_2C_2Se_2$	OC	1.775(9,4)	N,N	86.7(3)	
					N	2.002(7,7)	N,C	91.3(3,8)	
					Se	2.4448(14,54)		169.6(3,2)	
							C,C	92.5(4)	
							N,Se	72.0(2,1)[g]	
								96.0(2,9)	
							C,Se	95.6(3,3.0)	
							Se,Se	163.75(6)	
$[Fe(dppe)(SPh)_2(CO)_2]$[f]	tr	11.490(10)	85.32(3)	$FeC_2S_2P_2$	OC	1.753(15,17)	C,C	90.6(7)	138
(red)	P-1	16.666(8)	97.17(6)		PhS	2.331(5,8)	C,P	92.9(5,2.0)	
	4	19.387(5)	105.32(8)		dppeP	2.263(5,31)		179.0(5)	
							C,S	94.4(5,4.3)	
								175.5(6)	
							P,S	87.7(2,1.7)	
								172.7(2)	
							P,P	86.2(1)[d]	
							S,S	87.2(2)	
				$FeC_2S_2P_2$	OC	1.769(16,13)	C,C	90.6(7)	
					PhS	2.348(5,2)	C,P	93.9(5,3.0)	
					dppeP	2.265(4,16)		176.0(4)	
							C,S	90.2(4,3.9)	
								177.2(5)	
							P,S	87.6(2,2.4)	
								173.3(2)	
							P,P	86.7(1)[d]	
							S,S	87.4(2)	

Table 3. Continued 13

Compound	Cryst. class / Space group / Z	a, b, c	α, β, γ	Chromophore	Fe–L	L–Fe–L	Ref.
cct-[Fe(PMe$_3$)$_2$(SeMe)$_2$(CO)$_2$][f]	m	14.597(2)		FeC$_2$P$_2$Se$_2$	OC 1.745(7,6)	C,C 95.6(3)	139
(red-brown)	P2$_1$/n	16.978(2)	108.63(2)		Me$_3$P 2.257(2,3)	C,P 91.4(2,1.4)	
	8	15.317(4)			MeSe 2.448(1,1)	C,Se 90.7(2,1.60	
						173.1(2,1.5)	
						P,P 175.53(9)	
						P,Se 88.47(7,4.31)	
						Se,Se 83.25(4)	
				FeC$_2$P$_2$Se$_2$	OC 1.751(7,7)	C,C 96.5(3)	
					Me$_3$P 2.263(1,1)	C,P 91.3(2,1.6)	
					MeSe 2.449(1,8)	C,Se 90.3(2,3)	
						172.7(2,3)	
						P,P 175.90(8)	
						P,Se 88.49(6,3.44)	
						Se,Se 83.05(4)	
[Fe(PPh$_3$)(SiPh$_3$)(CO)$_3$(H)]	tr	12.32(2)	106.3(2)	FeC$_3$HPSi	OC 1.761(46,37)	C,C 100.7(20,9)	140a
(not given)	P-1	14.50(4)	98.2(2)		H not located	158.6(17)	
(at 223 K)	2	10.06(2)	72.0(2)		Ph$_3$P 2.271(12)	C,P 92.2(15,4.7)	
					Ph$_3$Si 2.343(46)	C,Si 90.4(15,4.7)	
						P,Si 168.0(4)	
[Fe{P(OPh)$_3$}(SiPh$_3$).	tr	11.509(2)	108.04(2)	FeC$_3$HPSi	OC 1.775(3,19)	C,C 103.1(1,7)	140b
(CO)$_3$(H)]	P-1	14.403(3)	95.10(2)		H 1.46(3)	151.0(1)	
(not given)	2	11.153(4)	92.55(2)		(PhO)$_3$P 2.163(1)	C,H 76(1,2)	
					Ph$_3$Si 2.380(1)	174(1)	
						C,P 95.34(9,1.33)	
						C,Si 85.94(9,4.49)	
						H,P 88(1)	
						H,Si 85(1)	
						P,Si 172.65(3)	

Table 3. Continued 14

[Fe(NP$_3$)(CO)(H)](BPh$_4$) (not given)	m P2$_1$/c 4	19.284(4) 18.029(4) 16.936(3)	113.73(3)	FeP$_3$HNC	P 2.210(3,0) N 2.126(6) OC 1.735(8) H 1.43(9)	P,P 108.3(1,2) 142.1(1) P.N 85.9(2) 86.3(2,8)[d] P,C 93.9(3,1.9) P,H 71(4,8) 168(4) N,C 176.7(3) C,H 94(3)	141
cct-[Fe{P(OPh)$_3$}$_2$(SPh). (CO)$_2$(H)] (orange)	m P2$_1$/n 4	9.935(2) 36.185(5) 12.112(2)	113.42(1)	FeC$_2$P$_2$HS	OC 1.779(8,9) P 2.157(3,5) H 1.41(7) PhS 2.343(3)	C,C 95.2(5,0) C,P 97.6(3,4.3) C,H 81(4) C,S 95.1(4) 169.3(3) P,P 163.2(1) P,H 83(3,8) P,S not given H,S 89(4)	142
trans-[Fe(PPh$_3$)$_2$(CO)$_2$(H)I] (not given)	m C2/c 4	23.360(11) 9.783(5) 22.257(11)	138.55(2)	FeC$_2$P$_2$HI	OC 1.718(8,22) Ph$_3$P 2.252(1) H 1.37(11) I 2.635(7)	C,H 79.5(3) P,H 85.59(4) C,I 100.5(3) P,I 94.41(4)	143
[Fe(η^2-COPr^i)(PEt$_3$)$_2$)(CO)I] (red brown) (at 238 K)	m P2$_2$/c 8	19.06(3) 13.09(2) 22.83(4)	128.4(1)	FeC$_2$P$_2$OI	OC not given Pr^{i}OC 1.80(2) Pr^{i}CO 2.19(1) Et$_3$P not given I not given	OC,I 115.6(6)	144

a Where more than one chemically equivalent distance or angles is present, the mean value is tabulated. The first number in parenthesis is e.s.d. and second is maximum deviation from the mean value.
b The chemical identity of coordinated atom/ligand is specified in these columns.
c Six-membered metallocyclic ring.
d Five-membered metallocyclic ring.
e Three-membered metallocyclic ring.
f There are two crystallographically independent molecules.
g Four-membered metallocyclic ring.

Table 4. Crystallographic and Structural Data of Iron Carbonyls with Coordination Number Seven and Eight[a]

COMPOUND (COLOUR)	CRYST. CL. SPACE GR. Z	a[Å] b[Å] c[Å]	α[°] β[°] γ[°]	CHROMO-PHORE	Fe–L [Å]	L–Fe–L [°]	REF.
$(Ph_4P)_2[Fe(As_3Se_3)_2(CO)]$ (brown)	tr P-1 2	13.114(2) 19.250(2) 11.791(2)	98.85(1) 108.81(2) 88.86(1)	$FeAs_4Se_2C$	As[b] 2.442(4,4) Se 2.422(4,1) OC 1.691(23)	As,As[b] 58.3(1,1)[c] 78.6(1) 136.7(1,10 164.9(2) Se,Se 178.3(2) As,Se 81.3(1,1.3) 98.9(1,7)[d] As,C 82.5(7,1.4) 140.6(7,1.1) Se,C 89.7(7,3.6)	145
$(NBu^n{}_4)[Fe(B_5H_7)(CO)_3(H)]$ (red)	m $P2_1/b$ 4	11.108(6) 15.462(12) 15.443(9)	 95.87(5)	FeC_3B_3H	OC not given B 2.12(-,4) H 1.52	B,B 83.5(6)	146
$[Fe\{\eta^4\text{-}(PrBNBu^t)_2\}(CO)_3]$ (not given)	or Pbcn 8	31.077(12) 8.814(2) 16.061(7)		$FeC_3N_2B_2$	OC 1.763(7,21) N 2.085(5,19) B 2.189(7,9)	C,C 90.1(3,1.9) 114.5(3)	147
$[Fe(\eta^4\text{-}(mesBPBu^t)_2\}(CO)_3]$ (light yellow)	m $P2_1/c$ 4	11.722(6) 16.259(4) 18.37(1)	 108.24(4)	$FeC_3B_2P_2$	OC 1.773(10,2) B 2.332(9,7) P 2.292(3,3)	C,C not given B,B 74.7(3) P,P 67.6(1) B,P 48.6(2,9)	148
$[Fe(C_2B_9H_{11})(CO)_3]$ (golden yellow)	m $P2_2/n$ 4	6.971(4) 15.900(9) 11.237(7)	 91.42(2)	FeC_5B_3	OC 1.804(11,40) C 2.110(8,4) B 2.167(9,7)	C,C 45.46(28)[c] 90.7- 106.39(35) 156.05(34, 4.34) B,B 49.41(34,25)[c] 84.17(34) C,B 46.48(32,5)[c] 78.77-99.22(36 124.49(39,5.32)	149

Table 4. Continued 2

Compound	Cryst. cl. / Space grp. / Z	a [Å] / b [Å] / c [Å]	β [°]	Chromophore	Fe-L [Å]	L-Fe-L [°]	Ref.
$[Fe(PPh_3)(C_2B_9H_{11})(CO)_2]$	m	18.384(3)		FeC_4B_3P	OC 1.758(12,4)	C,C 45.20(38)[c]	149
(yellow)	A2/a	12.762(2)	104.081(4)		C 2.109(10,4)	91.32(57)	
	8	23.059(3)			B 2.190(13,29)	111.7(5,5.3)	
					Ph_3P 2.293(3)	155.9(5,4.3)49	
						B,B 49.4(4,1)[c]	
						83.97(48)	
						C,B 46.5(4,3)[c]	
						79.44-	
						99.37(49)	
						139.9(5,7.4)	
						C,P 92.8(5,6.0)	
						B,P 128.4(4,7.8)	
						170.0(33)	
$[Fe\{P(OMe)_3\}_2(C_2B_9H_{11})(CO)]$	or	10.397(2)		$FeC_3B_3P_2$	OC 1.770(14)	C,C 45.7(4)[c]	149
(yellow)	Pbnm	14.419(3)			C 2.109(14,13)	79.2(4)	
	4	15.092(3)			B 2.176(12)	125.4(4)	
					$(MeO)_3P$ 2.181(3)	163.8(4)	
						B,B 48.7(5)[c]	
						P,P 89.7(1)	
						C,B 47.5(4)[c]	
						83.0(4,2.1)	
						C,P 84.4(3)	
						95.7)3.4.9)	
						143.2(3)	
						B,P 110.7(3)	
						159.2(3)	

Table 4. Continued 3

[Fe(PPh_3)(MeCN)($C_2B_9H_{11}$).	m	28.050(2)		FeC_3B_3NP	OC 1.749(11)	C,C	44.70(34)	149
(CO)]	C2/c	11.571(9)	116.846(2)		C 2.111(9,8)		126.97(43)	
(red)	8	19.042(2)			B 2.157(11,35)		161.83(41)	
					MeCN 1.935(9)	B,B	50.04(44,11)	
					Ph_3P 2.265(3)		85.27(46)	
						C,N	93.0(4,7.6)	
						C,P	95.9(3,4.1)	
							140.79(29)	
						C,B	47.6(4,1.0)	
							81.8(4,2.8)	
							119.95(45)	
						B,N	107.97(40)	
							151.23(40,6.22)	
						B,P	84.24(32)	
							115.26(32)	
							165.35(34)	
						N,P	86.41(25)	

a Where more than one chemically equivalent distance or angles is present, the mean value is tabulated. The first number in parenthesis is e.s.d, and second is maximum deviation from tne mean value.
b The chemical identity of coordinated atom/ligand is specified in these columns.
c Three-membered metallocyclic ring .
d Five-membered metallocyclic ring.

Table 5. Crystallographic and Structural Data for Binuclear Iron Carbonyl Compounds[a]

COMPOUND (colour)	CRYST. CL. SPACE G. Z	a [Å] b [Å] c [Å]	α [°] β [°] γ [°]	CHROMO-PHORE	Fe-L [Å]	Fe-Fe [Å] Fe-L-Fe [°]	L-Fe-L [°]	REF.
$[(OC)_6Fe_2\{\mu\text{-dhs}\}]$ (red)	tr P1 2	8.59(2) 13.92(3) 7.85(2)	92.14(15) 93.24(15) 90.24(15)	FeC_3N_2	OC[b] 1.784(16,38) μN 2.005(7,26)	2.372(2) N[b] 72.5(4,7)	C,C[b] 95.0(6,5.4) N,N 72.9(4,1)	150
$[(OC)_6Fe_2\{\mu\text{-PhNSO}_2\text{-NPh}\}]$ (orange)	or Cmcm 4	11.705(2) 10.816(2) 15.950(3)		FeC_3N_2	OC not given μN 2.005(3,0)	2.378(1) N 72.7(1)	N,N 67.1(2)	151
$[(OC)_6Fe_2\{\mu\text{-N(H)C}_6H_4N\text{-(CMe}_3)\}]$ (red)	tr P-1 2	8.029(2) 8.676(3) 14.289(4)	83.84(3) 79.49(2) 71.96(3)	FeC_3N_2	OC 1.782(8,40) μN 2.014(6,52)	2.384(1) N not given	C,C 92.1(4,6.6) N,N 72.9(2,1) C,N 98.0(3,7.4) 156.1(3)	152
$[(OC)_6Fe_2\{\mu\text{-(MeN)}_2CO\}]$ (orange)	tr P-1 2	7.79(2) 7.87(2) 11.88(2)	92.18(10) 103.45(10) 97.38(10)	FeC_3N_2	OC 1.76(4,7) μN 1.97(2,2)	2.391(7) N 75.0(7,7)	C,C 95.9(13,2.4) N,N 64.8(8,3) C,N 101.4(11,6.7) 152.9(11,1.7)	153
$[(OC)_6Fe_2\{\mu\text{-PhNC(O)}.C_6H_4NH\}]$ (bright red) (at 173 K)	or Pbca 8	14.454(2) 12.453(2) 21.167(2)		FeC_3N_2	OC not given μN 1.970(2,2) 2.055(2,3)	2.402(1) N 73.31(8,21)	N,N 79.28(9,3)	154
$[(OC)_6Fe_2.\{\mu\text{-}\eta^2\text{-C}_6H_4SNH\}]$ (not given)	m $P2_1/c$ 4	8.982(3) 14.443(5) 11.597(5)	94.39(5)	FeC_3NS	OC 1.788(8,44) μN 2.002(5,3) μS 2.283(2,5)	2.411(1) N 74.03(19) S 63.73(5)	C,C 96.6(4,7.5) N,S 77.4(1,1)	155
$[(OC)_3Fe\{\mu\text{-NCHMe}\}]_2$ (yellow orange)	or Pnma 4	7.106(3) 15.062(4) 14.180(4)		FeC_3N_2	OC 1.798(16,15) μN 1.942(7,4)	2.415(2) N 76.9(2)	C,C 95.7(5,4.3) N,N 76.0(3) C,N 97.4(4,6.9) 156.5(4,5)	156

Table 5. Continued 2

Compound				Chromophore	Fe–L (Å)	Fe–Fe (Å); Fe–X–Fe (°)	L–Fe–L (°)	Ref.
[(OC)$_6$Fe$_2${μ-dmsp}] (not given) (at 120 K)	m P2$_1$/c 4	8.492(1) 15.737(1) 15.499(1)	 125.129(5)	FeC$_3$OS	OC 1.803(3,13) μO 1.964(2,0) μS 2.2487(6,19)	2.4152(4) O 75.87(4) S 64.96(2)	C,C 96.6(1,7.2) C,O 98.5(1,5.3) 158.0(1,1.0) C,S 98.8(1,9.2) O,S 76.19(5,4)	157
[(OC)$_6$Fe$_2${μ-4-MeC$_6$H$_4$NC-(O)S}] (not given)	m P2$_1$/n 4	7.301(1) 29.845(2) 7.960(1)	 90.20(1)	FeC$_3$NS	OC 1.82(2,6) μN 2.02(1,0) μS 2.274(5,5)	2.431(4) N 74.0(4) S 64.6(1)	C,C 96(1,6) C,N 155(1,1) C,S 157(1,1) N,S 70.9(4,1)	158
[(OC)$_6$Fe$_2${μ-NC$_6$H$_{10}$}$_2$] (not given)	m P2$_1$/n 4	8.849(1) 27.636(6) 9.545(1)	 114.11	FeC$_3$N$_2$	OC 1.79(1,3) μN 1.92(1,1)	2.438(3) N 79.1(4,1)	C,C 97.3(8,4.8) N,N 78.1(8,2) C,N 96.6(8,8.6) 156.4(8,2.1)	159

Table 5. Continued 3

Compound (colour)	Cryst. cl. / Space group / Z	a / b / c	β	Chromophore	Fe–L	Fe–Fe / Fe–L–Fe	L–Fe–L	Ref.
$[(OC)_6Fe_2\{\mu\text{-}N(Me)Ph\}\text{-}$ $\{\mu\text{-}OCN(Me)Ph\}]^c$ (red)	m $P2_1/n$ 8	19.919(7) 10.122(4) 22.548(9)	98.90(3)	FeC_4N	OC 1.796(7,19) LC 1.983(5) μN 2.015(5)	2.461(14) N 75.4(2)	C,C 93.6(3,5.8) 97.0(2,2.9) 168.1(3) C,N 101.4(2,7.3) 151.6(2)	160
				FeC_3ON	OC 1.806(8,40) LO 1.978(5) μN 2.007(5)		C,C 92.8(3,4.9) C,O 93.2(2,5.1) 169.4(2) C,N 102.3(2,6.9) 152.2(3) O,N 83.6(2)	
				FeC_4N	OC 1.805(7,14) LC 1.976(5) μN 2.015(5)	2.477(14) N 76.1(2)	C,C 93.0(3,4.6) 93.8(2,6.3) 168.2(3) C,N 101.7(3,6.8) 152.1(2)	
				FeC_3ON	OC 1.792(8,32) LO 1.966(4) μN 2.003(5)		C,C 93.6(4,3.8) C,O 90.9(3,4.6) 170.1(3) C,N 101.7(3,7.1) 154.3(2) O,N 84.4(2)	
$[(OC)_6Fe_2\{\mu\text{-}tcgd\}]$ (orange)	or Pnma 4	9.088(2) 10.381(6) 17.035(9)		FeC_3S_2	OC 1.74(4,4) μS 2.261(4,0)	2.465(4) S 66.1(1)	C,C 96(1,3) S,S 82.5(2,0) C,S 96.4(8,11.7) 156.5(8,1.0)	161
$[(OC)_6Fe_2\{\mu\text{-}bcgdt\}]$ (orange)	or Pnma 4	9.308(3) 10.544(3) 16.890(6)		FeC_3S_2	OC 1.80(1,3) μS 2.254(3,3)	2.469(3) S 66.4(1)	C,C 95.3(6,6.2) S,S 97.4(1,1) C,S 98.1(4,9.9) 156.1(4,2.6)	161

Table 5. Continued 4

Compound (colour)	Cryst. cl. Space group Z	a [Å] b [Å] c [Å]	α [°] β [°] γ [°]	Chromophore	Fe-L [Å]	Fe-Fe [Å] Fe-X-Fe [°]	L-Fe-L [°]	Ref.
[(OC)$_6$Fe$_2${μ-fdt}] (red orange)	m P2$_1$/c 4	9.510(6) 18.425(9) 8.935(5)	 107.61(3)	FeC$_3$S$_2$	OC 1.79(1,5) μS 2.264(3,3)	2.483(2) S 66.5(1,0)	C,C 95.9(5,5.1) S,S 72.8(1,1) C,S 98.7(5,6.8) 156.7(4,1.1)	162
[(OC)$_6$Fe$_2${μ-NCMe$_2$}- {μ-η^2-ONCMe$_2$} (orange)	tr P-1 2	12.061(10) 9.119(7) 8.599(7)	66.81(2) 99.94(3) 107.08(2)	FeC$_3$N$_2$	OC 1.791(7,15) μN 1.921(4) N 1.982(5)	2.484(2) N 81.1(2)	C,C 95.8(3,6.9) N,N 85.6(2) C,N 95.4(2,10.7) 159.7(2,8.6)	163
				FeC$_3$ON	OC 1.793(7,31) O 1.952(4) μN 1.902(4)		C,C 95.2(3,2.9) C,O 90.3(2,1.5) 173.3(2) C,N 100.5(2,8.5) 151.6(2) O,N 83.9(2)	
[(OC)$_6$Fe$_2${μ-MeC(O)- CH$_2$CHS$_2$}] (red)	m C2/c 8	15.378(2) 9.6620(9) 21.768(3)	 108.23(1)	FeC$_3$S$_2$	OC not given μS 2.265(2,11)	2.485(1) S not given	not given	164
[(OC)$_6$Fe$_2${μ-mdt}] (not given)	m P2$_1$/n 4	11.752(3) 12.038(2) 8.808(2)	 100.75(2)	FeC$_3$S$_2$	OC not given μS 2.262(1)	2.485(1) S 66.64(4)	S,S 72.45(4)	165
[(OC)$_6$Fe$_2${μ-MeC(O)- CH$_2$CHS$_2$}] (red)	m C2/c 8	15.378(2) 9.6620(9) 21.768(3)	 108.23(1)	FeC$_3$S$_2$	OC not given μS 2.265(2,11)	2.485(1) S not given	not given	166
[(OC)$_5$Fe$_2${μ-P(Bu^t)$_2$}$_2$] (yellow)	tr P-1 2	11.741(1) 13.662(2) 8.652(1)	101.21(1) 103.64(1) 76.63(1)	FeC$_3$P$_2$	OC not given μP 2.3687(8)	2.4870(6) P not given	not given	167
[(OC)$_6$Fe$_2${μ-S$_2$CO}] (not given)	or Pnam 4	14.131(7) 6.764(3) 12.900(8)		FeC$_3$S$_2$	OC 1.802(6,11) μS 2.268(2,2)	2.4875(9) S 66.50(5,5)	C,C 96.0(3,5.2) S,S 75.16(6) C,S 97.6(2,6.2) 158.(2,1.4)	168

Table 5. Continued 5

[(OC)$_6$Fe$_2${μ-htcgd}] (orange)	or Pnam 4	9.158(3) 10.165(4) 17.057(5)		FeC$_3$S$_2$	OC 1.792(8,14) μS 2.262(1,1)	2.489(1) S 66.8(1)	C,C 95.7(4,3.5) S,S 83.4(1,1) C,S 94.4(2,11.5) 155.9(2,6)	169
[(OC)$_6$Fe$_2${μ-Me$_3$SiCH- (CH$_2$)S$_2$}] (not given)	tr P-1 2	9.304(3) 9.418(3) 11.646(3)	114.2(3) 92.21(2) 101.75(2)	FeC$_3$S$_2$	OC 1.803(-,55) μS 2.252(-,7)	2.492(-) 67.2(-,0)	not given	170
[(OC)$_6$Fe$_2${μ-Ph$_2$N$_2$}] (red)	m P2$_1$/n 4	7.906(2) 16.811(3) 14.788(3)	99.56(3)	FeC$_3$N$_2$	OC 1.804(4,8) μN 1.908(3,15)	2.493(1) N 81.6(1,0)	N,N 43.3(1,2)	171
[(OC)$_6$Fe$_2${μ-hdt}] (orange)	tr P-1 2	10.152(6) 9.385(6) 9.110(6)	96.95(4) 89.51(4) 107.15(4)	FeC$_3$S$_2$	OC 1.74(2,5) μS 2.256(6,13)	2.493(4) S 67.1(2,0)	C,C 96(1,5) S,S 71.1(2,3) C,S 99.0(7,6.4) 155.5(8,1.7)	172
[N(H)Et$_3$][(OC)$_6$Fe$_2$- {μ-CO}{μ-C$_6$H$_2$Me$_3$S}] (not given)				FeC$_4$S	OC 1.786(8,54) μOC 1.926(7,1) μS 2.304(2,2)	2.494(1) C 80.7(3) S 65.5(1)	C,C 93.3(3,9.8) 149.6(3,2.7) C,S 75.9(2,1) 97.2(3,2.6) 158.1(2,9)	173
[(OC)$_6$Fe$_2${μ-chdt}] (orange)	m P2$_1$/c 4	9.167(3) 14.722(7) 11.931(5)	95.95(3)	FeC$_3$S$_2$	OC 1.78(1,2) μS 2.242(3,6)	2.496(2) S 67.63(8,2)	C,C 97.0(5,4.8) S,S 80.19(9,20)	174
[(OC)$_6$Fe$_2${μ-SCH$_2$CH$_2$S}] (red)	tr P-1 4	8.961(3) 9.277(3) 16.023(6)	86.15(3) 85.18(3) 88.49(3)	FeC$_3$S$_2$	OC not given μS 2.245(2)	2.497(4) S 67.6(2)	S,S 80.2(1)	175
[(OC)$_6$Fe$_2${μ-1,3-C$_7$H$_8$S$_2$}] (dark red)	m P2$_1$/n 4	12.981(1) 13.294(1) 9.583(1)	98.46(1)	FeC$_3$S$_2$	not given	2.497(4) S not given	S,S 84.5(1)	176

Table 5. Continued 6

[(OC)$_6$Fe$_2${μ-η^2-dmda} (deep red)	tr P-1 2	7.523(1) 9.801(3) 11.825(1)	76.34(2) 72.87(1) 78.74(2)	FeC$_3$N$_2$	OC 1.799(4,10) μN 1.906(3,14)	2.498(1) N not given	C,C 97	177
[(OC)$_6$Fe$_2${μ-((CH$_2$NMe)$_2$-B$_2$)S$_2$}] (dark brown)	tr P-1 2	8.262(4) 11.276(3) 13.325(4)	112.17(2) 93.77(3) 99.33(3)	FeC$_3$S$_2$	OC not given μS 2.294(2,12)	2.498(3) S 66.0(1,0)	S,S 75.2(1,1)	178
[(OC)$_6$Fe$_2${μ-(Me$_2$N)$_3$PCO-CS$_2$}] (red)	or Pbca 8	15.131(3) 13.243(1) 22.784(2)		FeC$_3$S$_2$	OC 1.800(14,38) μS 2.254(4,10)	2.499(3) S 67.3(1,2)	C,C 97.5(7,5.0) S,S 79.4(1,1) C,S 96.0(5,9.6) 157.2(5,3.3)	179
[(OC)$_6$Fe$_2${μ-SCPh}$_2$] (not given)	tr P-1 2	35.68 9.30 28.39	90.15 108.43 103.20	FeC$_3$S$_2$	OC not given μS 2.26	2.50 S not given	not given	180
[(OC)$_6$Fe$_2${μ-SC(Ph).C(H)Se}] (deep red)	m P2$_1$/c 4	17.974(2) 6.822(10) 14.008(2)	97.070(10)	FeC$_3$SSe	OC not given μS 2.289(3,0) μSe 2.383(2,3)	2.501(3) S 66.2(1) Se 63.3(1)	S,Se 81.1(1,1)	181
[(OC)$_6$Fe$_2${μ-(Me$_2$NB)$_2$S$_2$}][c] (dark red)	m P2$_1$/n 8	12.578(7) 15.940(4) 19.198(9)	101.03(4)	FeC$_3$S$_2$	OC not given μS 2.295(2,6)	2.501(2) S 66.0(1,1)	S,S 85.4(1,2)	178
				FeC$_3$S$_2$	OC not given μS 2.289(2,5)	2.510(2) S 66.4(1,1)	S,S 85.2(1,1)	
[(OC)$_6$Fe$_2${μ-SCH$_2$}$_2$] (not given)	tr P-1 4	9.177(3) 15.861(3) 8.906(2)	94.93(2) 91.33(2) 85.77(2)	FeC$_3$S$_2$	OC not given μS 2.239(1,11)	2.502(1) S 68.12(3)	S,S 80.14(4)	182
[(OC)$_6$Fe$_2${μ-Bu$^t_3$(SiMe$_3$).N$_2$hx}] (red)	m C2/c 4	17.243(2) 10.600(1) 18.386(2)	117.80(1)	FeC$_3$N$_2$	OC not given μN 1.923(4,2)	2.502(2) N not given	N,C 110.2(3,3.7) 142.7(4)	183

Table 5. Continued 7

$[(OC)_6Fe_2\{\mu\text{-}S(OMe)CH_2\}.$ $\{\mu\text{-}Sbz\}]$ (dark red)	m $P2_1/n$ 4	9.023(3) 12.081(5) 16.720(11)	 97.04	FeC_3S_2	OC 1.794(8,26) μPhS 2.258(2,3) μLS 2.250(2,5)	2.504(1) S 67.6(1,2)	C,C S,S C,S	96.2(3,5.1) 79.8(6,0) 97.2(2,12.8) 156.3(2,5.5)	184
$[(OC)_6Fe_2\{\mu\text{-}SC_6H_4\text{-}4\text{-}$ $OH\}_2]\cdot C_6H_6$ (red)	tr P-1 2	10.305(5) 11.898(6) 12.037(6)	102.85(4) 95.47(4) 112.06(4)	FeC_3S_2	OC 1.77(2,5) μS 2.263(5,19)	2.504(3) S 67.2(2,3)	C,C S,S C,S	96.6(9,4.7) 79.6(2,3) 97.0(7,11.8) 156.3(7,3.2)	185
$[(OC)_6Fe_2\{\mu\text{-}tmpBS_2\}]$ (dark red)	m $P2_1/n$ 4	10.656(4) 14.678(4) 14.607(5)	 108.79(3)	FeC_3S_2	OC not given μS 2.274(2,4)	2.506(1) S 66.8(1,1)	S,S	76.2(1,0)	178
$[(OC)_6Fe_2\{\mu\text{-}Me_3Si(H).$ $(CH_2S)_2\}]$ (not given)	tr P1 2	7.625 11.800 12.260	115.62 93.78 96.93	FeC_3S	OC 1.776(-,69) μS 2.245(-,2)	2.506 S not given		not given	186
$[(OC)_6Fe_2\{\mu\text{-}Se_2(HC{=}C\text{-}$ $(C{\equiv}CMe))\}]$ (purple)	m $P2_1/n$ 4	6.999(2) 18.701(6) 12.065(3)	 104.19(2)	FeC_3Se_2	OC not given μSe 2.384(2,3)	2.507(2) Se 63.5(1,1)	Se,Se	81.9(1,0)	187
$[(OC)_6Fe_2\{\mu\text{-}S_2C_{60}\}]$ (not given) (at 198(2) K)	trg R3m 9	27.284(8) 14.675(3)		FeC_3S_2	OC 1.817(7,6) μS 2.232(2,5)	2.510(2) S 68.4(1,2)	S,S C,S	80.4(7) 100.0(2,9.0) 152.7(2)	188
$[(OC)_6Fe_2\{\mu\text{-}OH\}.$ $\{\mu\text{-}P(p\text{-}C_6H_4Me)_2\}]$ (orange red)	m C2/c 8	10.534(6) 17.231(8) 23.810(15)	 100.72(5)	FeC_3OP	OC 1.795(14,31) μHO 1.972(6,3) μP 2.238(3,2)	2.511(2) O 79.1(2) P 68.3(1)	C,C C,O O,P	95.5(5,7.0) 97.0(4,5.6) 163.4(4,2) 78.5(2,1)	189
$[(OC)_6Fe_2\{\mu\text{-}S\}\{\mu\text{-}SMe)]\cdot$ C_4H_8 (dark red)	m $P2_1/a$ 2	7.910(1) 12.564(2) 14.644(2)	 101.28(1)	FeC_3S_2	OC not given μS 2.249(2,2) μMeS 2.263(2,2)	2.511(2) S not given		not given	190
$[(OC)_6Fe_2\{\mu\text{-}SeC(Ph).$ $C(H)Se\}]$ (purple)	m $P2_1/a$ 4	14.561(1) 6.666(1) 17.475(1)	 91.22(1)	FeC_3Se_2	OC not given μSe 2.3837(9,64)	2.512(1) Se 63.60(3,3)	Se,Se	81.37(3,5)	191

Table 5. Continued 8

$[(OC)_3Fe\{\mu\text{-}SPh\}]_2$ (red orange)	m C2/c 8	26.339(2) 11.664(1) 13.755(1)	106.705(5)	FeC_3S_2	OC 1.807(7,24) μS 2.270(2,11)	2.516(2) S 67.3(1,3)	C,C 96.8(3,7.2) S,S 79.8(1,1) C,S 98.6(2,9.8) 156.5(3,3.2)	192
$[\{(CO)_3Fe\}_2(\mu\text{-}MeCMeSi.\text{-}PH_2CMeCMe)]$ (orange)	m $P2_1/c$ 4	11.163(3) 16.191(2) 15.165(3)	109.02(2)	FeC_6	OC 1.785(6,10) μC 2.052(5,58) C 2.189(4,2)	2.5710(9) C not given	C,C 92.2(2,9.8) 169.9(2,2.4)	116
$ppn[(OC)_6Fe_2\{\mu\text{-}H\}\{\mu\text{-}CO\}_2]$ (yellow brown)	m $P2_1/c$ 4	10.667(5) 20.470(10) 19.421(10)	107.72(3)	FeC_5H	OC 1.788(11,32) μOC 1.970(12,11) μH 1.61(8,4)	2.521(1) C 79.6(5,2) H 103(4)	C,C 81.1(4,3) 94.8(5,8.7) 164.0(5,2.8) C,H 76(3,1) 89(3,4) 164(3,3)	193
$[(OC)_6Fe_2\{\mu\text{-}S_2Ph_4C_6\}]$[d] (brown black)	m $P2_1/c$ 12	13.297(2) 22.443(5) 35.228(9)	93.26(2)	FeC_3S_2	OC 1.737 μS 2.250(7,14)	2.521(4) S 68.1(2,0)	C,C 97(1,5) S,S 83.6(2,2)	194
				FeC_3S_2	OC 1.737 μS 2.242(6,6)	2.526(5) S 68.5(2,2)	C,C 97(1,6) S,S 83.1(2,1)	
				FeC_3S_2	OC 1.737 μS 2.242(6,9)	2.530(4) S 68.7(2,2)	C,C 97(1,4) S,S 83.6(2,1)	
$[(OC)_6Fe_2\{\mu\text{-}SC_6H_4\text{-}4\text{-}OH\}_2]\cdot 2HSC_6H_4\text{-}4\text{-}OH$ (red)	or $P2_12_12_1$ 4	8.967(4) 12.218(5) 31.708(8)		FeC_3S_2	OC 1.789(6,26) μS 2.267(1,19)	2.522(1) S 67.61(4,32)	C,C 96.1(2,6.2) S,S 80.05(5,34) C,S 98.11(2,10.61) 157.15(2,4.75)	185
$[(OC)_8Fe_2\{\mu\text{-}CO\}]$ (not given)	tg $P6_3/m$ 1	6.436(1) 16.123(2)		FeC_5	OC 1.835(2) μOC 2.013(3)	2.523(1) C 77.6(1)	C,C 90.0(1,6.1)	195

Table 5. Continued 9

[(OC)$_5$(PPh$_3$)Fe$_2${μ-SEt}$_2$] (dark red)	tr P-1 2	10.472(4) 11.329(2) 13.437(2)	80.34(2) 92.62(3) 114.46(2)	FeC$_3$S$_2$	OC 1.776(5,20) μS 2.265(1,3)	2.524(9) S 67.7(1,3)	C,C 97.0(2,5.5) S,S 80.06(5) C,S 96.4(1,9.8) 156.2(1,4)	196
				FeC$_2$S$_2$P	OC 1.760(3,2) μS 2.268(1,13) Ph$_3$P 2.239(1)		C,C 92.3(2) S,S 79.93(5) C,S 89.8(1,2.9) 157.3(1,1.1) C,P 96.4(1,1) S,P 105.63(6,42)	
[(OC)$_4$Fe$_2${μ-η^2-(SMe)$_2$. C$_2$S$_2$}(PPh$_3$)$_2$] (purple)	tr P-1 2	9.669(2) 13.477(3) 17.323(3)	101.1(2) 91.5(2) 110.3(2)	FeC$_2$S$_2$P	OC 1.755(8,16) μS 2.279(2,24) Ph$_3$P 2.252(2,4)	2.525(2) S 66.6(1,5)	C,C 92.6(3,5) S,S 78.9(1,4) C,S 90.0(2,2.3) 154.7(3,3.8) C,P 98.4(3,1.2) S,P 103.9(1,2.9)	88
[(OC)$_6$Fe$_2${μ-SC(Ph). C(H)Te}] (red)	tr P-1 2	7.4914(90 10.5901(18) 12.1323(8)	105.533(7) 102.250(7) 106.553(15)	FeC$_3$STe	OC not given μS 2.266(2,3) μTe 2.546(1,1)	2.526(2) S 67.76(7) Te 59.47(4)	S,Te 81.00(7,4)	181
[(OC)$_6$Fe$_2${μ-SeCH$_2$Se}] (orange)	m P2$_1$/n 4	8.8812(4) 12.306(1) 11.810(1)	 101.25(3)	FeC$_3$Se$_2$	OC 1.785(6,20) μSe 2.3803(8,25)	2.527(1) Se 64.15(3,8)	C,C 97.2(3,5.2) Se,Se 74.35(3,8) C,Se 96.8(2,6.1) 158.0(2,1.3)	197
[(OC)$_6$Fe$_2${μ-Se(CH$_2$)$_2$Se}] (orange)	m P2$_1$/m 2	6.8216(4) 13.545(2) 7.948(1)	 113.716(8)	FeC$_3$Se$_2$	OC 1.782(7,10) μSe 2.361(1,1)	not given Se not given	C,C 96.7(3,5.7) Se,Se 82.26(3) C,Se 95.4(2,6.1) 159.0(2,8)	197
(AsPh$_4$) [(OC)$_6$Fe$_2${μ-S}- {μ-SSBu^t}] (dark green)	tr P1 2	11.398(5) 13.109(6) 14.316(6)	102.14 109.11 106.03	FeC$_3$S$_2$	OC not given μS 2.282(1,6) μBu^{t}S 2.249(1,1)	2.530(1) S 67.3 68.5	not given	198

Table 5. Continued 10

$[(OC)_6Fe_2\{\mu\text{-}SBu^t\}_2]$ (not given)	m $P2_1/c$ 4	8.777(4) 27.770(6) 8.986(3)	115.13(3)	FeC_3S_2	OC 1.80(1,5) μS 2.265(3,19)	2.530(2) S 67.9(1,4)	C,C 95.8(6,6.5) S,S 80.1(1,1) C,S 97.1(4,13.6) 156.7(4,3.5)	199
$[(OC)_5(PPh_3)Fe_2\{\mu\text{-}SeCl.$ $(CH_2OH)CHSe\}]$ (red)	tr P-1 2	10.517(3) 12.216(3) 13.148(3)	64.19(2) 67.43(1) 68.82(3)	FeC_3Se_2	OC 1.78(2,4) μSe 2.397(3,1)	2.533(3) Se 63.83(8,2)	C,C 97.0(8,5.9) Se,Se 80.46(9) C,Se 95.9(6,6.8) 158.0(5,2.8)	200
				FeC_2Se_2P	OC 1.74(2,2) μSe 2.395(3,2) Ph_3P 2.264(3)		C,C 91.8(7) Se,Se 80.55(9) C,Se 89.3(6,1.6) 156.0(5,2.0) C,P 98.7(4,2.0) Se,P 104.88(12,4.14)	
$[(OC)_6Fe_2\{\mu\text{-}PhCN_2S_2\}]$ (orange brown)	m $P2_1/c$ 4	13.941(1) 12.508(1) 10.077(1)	102.61(1)	FeC_3S_2	OC 1.795(8,21) μS 2.225(2,10)	2.533(2) S 69.4(1,3)	C,C 97.1(3,5.2) S,S 82.4(1,1) C,S 95.5(3,11.0) 157.8(3,4.2)	201
$[(OC)_3Fe\{\mu\text{-}mpt\}]_2$ (red)	m $P2_1/c$ 4	12.922(2) 14.784(5) 13.607(2)	112.41(1)	FeC_3S_2	OC 1.794(5,18) μS 2.262(1,9)	2.534(1) S 65.6(1,2.4)	S,S 81.10(4,10)	202
$[(OC)_4Fe_2\{\mu\text{-}SBu^n\}\{\mu\text{-}$ $MeOCO(CH_2)_2S\}(PPh_3)_2]$	m C2/c 8	30.892(4) 12.454(2) 25.725(4)	91.46(1)	FeC_2S_2P	OC 1.752(5,18) μBu^nS 2.274(1,3) μS 2.278(1,5) Ph_3P 2.263(1,15)	2.5364(8) S 67.7(1,1)	C,C 89.5(3,1.4) S,S 75.1(4,0) C,S 93.5(1,1.6) 156.7(1,4.0) C,P 100.2(2,2.7) S,P 98.6(1,0)	203
$[(OC)_6Fe_2\{\mu\text{-}Se_2\}_2\{\mu\text{-}HC.$ $C(C{\equiv}CMe)\}]{\cdot}C_4H_8O$ (orange)	tr P-1 2	11.281(2) 11.952(2) 12.614(2)	88.69(2) 64.95(1) 67.10(2)	FeC_3Se_2	OC not given μSe 2.378(2,11)	2.537(2,8) Se 64.5(1,9)	Se,Se 81.0(1,4)	187

Table 5. Continued 11

$[(OC)_6Fe_2\{\mu\text{-dsf}\}]$	tr	9.167(1)	92.56(1)	FeC_3Se_2	OC	1.790(5,9)	2.539(1)	C,C	96.4(3,4.9)	204
(dark red)	P-1	11.630(1)	90.33(1)		μSe	2.361(1,6)	Se 65.0(2,1)	Se,Se	81.6(0,1)	
	4	20.164(2)	95.05(1)					C,Se	96.5(2,10.7)	
									157.0(2,3.6)	
$[(OC)_6Fe_2\{\mu\text{-SeC(H)}.$	tr	7.773(9)	104.99(10)	FeC_3SeTe	OC	1.78(3,6)	2.539(5)	C,C	96.77(13,4.47)	205
$C(Ph)Te\}]$	P-1	10.635(9)	102.86(11)		μSe	2.372(6,5)	Se 64.71(18)	Se,Te	82.18(20,12)	
(red)	2	12.010(10)	106.38(9)		μTe	2.541(5,2)	Te 59.95(13)	C,Se	96.58(11,8.92)	
									157.6(8,1.5)	
								C,Te	94.2(8,7.3)	
									160.2(8,2.6)	
$[(OC)_6Fe_2\{\mu\text{-SeCH}_2Te\}]$	m	6.695(2)		FeC_3SeTe	OC	1.79(2,2)	not given	C,C	97.2(6,5.2)	206
(orange)	$P2_1/n$	13.993(5)	103.03(2)		μSe	2.454(2,1)	Se 62.7	C,Te	97.2(4,8.8)	
	4	14.007(4)			μTe	2.507(1,5)	Te 61.24(4)		157.3(4,4.3)	
$[(OC)_4Fe_2\{\mu\text{-OH}\}.$	tr	11.9981(30)	72.380(17)	FeC_2P_2O	OC	1.764(9,6)	2.546(1)	P,P	152.2(1,5)	207
$\{\mu\text{-PPh}_2\}\{\mu\text{-dppm}\}]$	P-1	12.3020(25)	75.990(18)		μPh_2P	2.212(2,10)	O 68.7(1)	P,O	82.8(1,4.5)	
(orange)	2	16.8524(38)	63.780(16)		μdppm,P	2.245(2,5)	P 70.2(1)			
					μHO	2.256(4,32)				
$[(OC)_6Fe_2\{\mu\text{-Pcy}_2\}.$	tr	11.625(4)	109.45(2)	FeC_3P_2		not given	not given		not given	208
$\{\mu\text{-PC(SiMe}_3)_2\}]$	P-1	17.766(5)	108.21(2)							
(red)	4	18.290(5)	96.89(3)							
$[(OC)_6Fe_2\{\mu\text{-S}\}_2]$	tr	6.63(1)	83.20(10)	FeC_3S_2	OC	1.776(5,-)	2.552(2)	C,C	96.5(7,2.3)	209
(orange)	P-1	7.85(1)	76.09(10)		μS	2.228(2,-)	S 69.9(1)	S,S	53.5(1,0)	
	2	11.46(1)	78.25(10)					C,S	102.9(5,1.3)	
									151.0(4,1.2)	
$[(OC)_6Fe_2\{\mu\text{-PPhCl}_2\}]$	tr	9.360	115.6	FeC_3ClP	OC	1.77(1,3)	2.56(0)	C,P	97.2(3,6.4)	210
(red brown)	P-1	9.538	104.2		μCl	2.29(1,0)	Cl 68.0(1)		154.8(3)	
	2	10.844	93.9		μP	2.20(1,0)	P 71.1(1)	Cl,P	78.2(1,1)	

Table 5. Continued 12

[(OC)$_6$Fe$_2${μ-Cl}{μ-PPh$_2$}] (red brown)	or Cmc2$_1$ 4	10.560(3) 12.604(2) 14.928(3)		FeC$_3$ClP	OC 1.796(4,25) μCl 2.289(1) μP 2.238(1)	2.5607(5) Cl 68.0(0) P 69.8(0)	C,C 96.1(2,6.1) C,Cl 94.9(1,4.5) 163.4(1) C,P 98.6(1,5.5) 153.6(1) Cl,P 79.4(0)	211
[(CO)$_2$Fe$_2${μ-CO}$_2$(bmpa)$_2$] (not given)	tr P-1 2	8.11(2) 10.25(1) 25.52(2)	115.92(9) 99.27(14) 113.19(13)	FeC$_3$N	μCO 1.924(5,6) OC 1.738(5,1) N 2.152(5,1)	2.562(1) C not given	C,C 89.3(2,3)	212
[(OC)$_6$Fe$_2${μ-SeMe}$_2$] (dark red)	or PC2$_1$n 4	7.202(1) 10.544(1) 18.344(3)		FeC$_3$Se$_2$	OC 1.82(3,6) μSe 2.381(10,17)	2.563(5) Se 65.2(2,1)	C,C 97(3,7) Se,Se 77.2(2,0) C,Se 96(2,8) 159(2,5)	213
[(OC)$_6$Fe$_2${μ-EtS}- {μ-p-MeC$_6$H$_4$SO$_2$}] (red)	m P2$_1$/a 4	16.027(4) 7.345(2) 16.665(3)	90.33(2)	FeC$_3$S$_2$	OC 1.796(7,18) μS 2.233(2,9)	2.568(1) S 69.80(4)	S,S 84.16(5) C,S 88.0(2) 155.2(2)	214
				FeC$_3$OS	OC 1.795(6,40) O 2.008(3) S 2.248(2)		C,S 106.8(2) C,O 89.0(2) O,S 83.7(1)	
[(OC)$_6$Fe$_2${μ-SC$_6$H$_{11}$}. {μ-PPh$_2$}] (orange red) (at 238 K)	m P2$_1$/c 4	15.071(6) 10.623(4) 16.259(6)	99.05(3)	FeC$_3$SP	OC 1.792(6,21) μS 2.271(2,1) μP 2.223(2,0)	2.568(2) S not given P 70.6(1)	C,C 94.9(3,5.3) S,P 76.5(1,0)	215
[(OC)$_6$Fe$_2${μ-CO}- {μ-C$_6$H$_4$N$_2$}] (blue green)	tr P-1 2	9.128(5) 10.574(6) 8.905(5)	95.26(5) 119.89(4) 100.46(4)	FeC$_4$N	OC 1.793(8,47) μOC 1.978(7,10) N 1.956(4,1)	2.573(1) C 81.4(2)	C,C 96.1(3,13.1) 96.9(3,11.2) 146.2(3,1.8) C,N 85.8(2,2) 94.2(3,2.0) 170.0(3,1)	216

Table 5. Continued 13

[(OC)$_6$Fe$_2${μ-C$_{31}$H$_{48}$P$_2$}] (not given)	or Pnma 4	9.846(2) 26.80(1) 15.112(4)		FeC$_3$P$_2$	OC not given μP 2.254(2,4)	2.574(2) P 69.63(6)	P,P 65.84(6,14)	217
[(OC)$_5$Fe$_2${μ-SPh}- {μ-PPh$_2$}(PPh$_3$)] (amber)	tr P-1 2	10.325(4) 10.890(5) 19.626(8)	69.54(4) 71.05(4) 67.85(4)	FeC$_3$SP	OC 1.791(10,17) μS 2.282(2) μP 2.245(3)	2.574(1) S not given P not given	C,C 97.6(5,6.7) C,S 99.2(3,5.8) 153.3(3) C,P 93.7(3,5.4) 158.7(3) S,P 78.3(1)	218
				FeC$_2$P$_2$S	OC 1.735(12,9) Ph$_3$P 2.279(2) μP 2.231(3) μS 2.285(2)		C,C 87.2(5) P,P 108.4(1) C,P 98.7(4,7.7) 144.9(3) C,S 91.6(3) 164.0(3) P,S 88.8(1,10.3)	
[(OC)$_6$Fe$_2${μ-(2,4,6- Pr^{i}C$_6$H$_2$P)$_2$CH$_2$}]	or Pnma 4	9.846(2) 26.80(1) 15.112(4)		FeC$_3$P$_2$	OC 1.779(10,2) μP 2.254(2,4)	2.574(2) P 69.63(6)	P,P 65.8(1,1)	219
[(OC)$_6$Fe$_2${μ-η^2-Se$_2$}] (red) (at 178(5) K)	tr P-1 2	7.704(2) 11.779(6) 6.585(3)	104.86(4) 102.28(3) 84.27(3)	FeC$_3$Se$_2$	OC 1.79(1,2) μSe 2.363(2,15)	2.575(2) Se 66.1(1,3)	Se,Se 58.1(1,1)	220
[(OC)$_4$Fe$_2${μ-H}{μ-CO}{μ- Pcy$_2${μ-dppm}] (orange)	tr P-1 2	10.2577(34) 13.0315(81) 15.8986(71)	81.902(43) 84.672(31) 72.272(38)	FeC$_3$P$_3$H	μOC 1.975(8,9) OC 1.790(9,30) μP 2.250(3,9) μdppm,P 2.245(3,12) μH not given	2.579(2) P 69.9(1)	C,C 166.9(3,1) P,P 152.7(1,1) C,P 76.6(3,1)	221
[(OC)$_4$Fe$_2${μ-I}{μ-Pcy$_2$}{μ- dppm}]·CH$_2$Cl$_2$ (orange)	tr P-1 2	11.8975(20) 12.9829(16) 16.6072(25)	72.937(12) 76.891(13) 64.089(11)	FeC$_2$P$_2$I	OC 1.749(12,35) μP 2.221(3,3) μdppm,P 2.240(3,5) μI 2.206(2,1)	2.580(2) P 71.0(1) I 59.4(1)	P,P 151.1(1) P,I 82.3(1,1)	221

Table 5. Continued 14

[(OC)$_6$Fe$_2${μ-P(Pr^i). Te-2,4,6-Me$_3$C$_6$H$_2$}. {μ-Te-2,4,6-Me$_3$C$_6$H$_2$}]	rh R-3c 12	25.53(2) 25.54(2) 25.56(1)	54.28(4) 54.24(4) 54.25(5)	FeC$_3$PTe	OC not given μP 2.243(10,8) μTe 2.583(4,4)	2.579(4) P 70.7(3) Te 60.4(1)	not given	222
[(OC)$_6$Fe$_2${μ- (Pr$^i_2$N)P- CH$_2$COCH$_2$P(NPr$^i_2$)}] (yellow)	or Pna2$_1$ 4	16.230(2) 20.700(2) 8.288(1)		FeC$_3$P$_2$	OC 1.788(5,8) μP 2.222(1,11)	2.579(1) P 70.9(1,0)	C,C 98.3(3,8.9) P,P 79.2(1,3) C,P 96.6(1,10.0) 153.9(1,10.1)	223
[(OC)$_6$Fe$_2${μ-TePhCHTe}] (orange red	m P2$_1$/n 4	8.254(3) 22.432(8) 9.078(2)	95.01(3)	FeC$_3$Te$_2$	OC 1.82(4,5) μTe 2.547(5,12)	2.585(6) Te 60.98(15,17)	Te,Te 75.72(14,2)	224
[(OC)$_6$Fe$_2${μ-SBu^t}. {μ-PC(SiMe$_3$)$_2$}] (orange)	m P2$_1$/n 4	11.002(2) 15.621(3) 15.772(2)	99.13(1)	FeC$_3$SP	OC 1.78(1,11) μS 2.270(3,10) μP 2.173(3,17)	2.586(2) S 69.42(8) P 73.0(1)	C,C 96.8(5,5.8) C,S 93.8(4,11.4) 159.0(4,3.9) C,P 98.2(4,9.6) 150.3(4,5) S,P 84.9(1)	225
[(OC)$_6$Fe$_2${μ-η^2- (Pr$^i_2$NP)$_2$CH$_2$CHCN}] (yellow)	tr P-1 2	8.783(2) 10.732(2) 15.753(2)	70.99(2) 77.00(2) 85.47(2)	FeC$_3$P$_2$	OC 1.76(1,4) μP 2.216(2,12)	2.587(2) P 71.40(7,9)	C,C 98.6(4,9.9) P,P 74.28(8,32) C,P 96.9(3,8.8) 148.0(3,1.1)	226
[(OC)$_6$Fe$_2${μ-η^2-(Pr$^i_2$NP)$_3$}] (deep orange)	m P2$_1$/a 4	14.910(5) 12.064(6) 19.733(10)	105.31(4)	FeC$_3$P$_2$	OC not given μP 2.243(4)	2.587(2) P not given	not given	227
[(OC)$_6$Fe$_2${μ-TeCH$_2$Te}] (dark red)	m P2$_1$/n 4	6.864(2) 13.926(3) 14.159(3)	104.43(2)	FeC$_3$Te$_2$	OC 1.782(16,40) μTe 2.549(2,3)	2.587(2) Te 61.1(1,1)	C,C 97.5(7,5.9) Te,Te 75.4(1,0) C,Te 96.7(5,8.3) 157.5(5,4.6)	228
[(OC)$_6$Fe$_2${μ-PMe$_2$}{μ-I}] (not given)	tr P-1 2	7.980(2) 12.473(2) 7.894(2)	101.22(1) 109.37(2) 81.19(2)	FeC$_3$PI	OC 1.79(1,3) μP 2.216(3,1) μI 2.598(2,7)	2.588(2) P 71.5(1) I 59.74(5)	P,I 83.5(1,1)	229

Table 5. Continued 15

[(OC)$_6$Fe$_2${μ-CNEt$_2$}(NO)] (red brown)	or Pbca 8	13.573(5) 14.675(2) 16.313(20)		FeC$_3$N	OC 1.742(16,25) μC 1.874(11) ON 1.686(12)	2.589(2) C 85.2(5)	C,C 96.3(7) NC,C 95.6(5) 138.7(6) ON,C 111.2(5,4.6) ON,NC 103.3(5)	
				FeC$_5$	OC 1.789(15,25) μC 1.950(10)		C,C 95.6(6,9.0) 168.8(6) NC,C 84.9(5,1.1) 109.6(5) 145.8(5)	160
[(OC)$_6$Fe$_2${μ-η^2-ac}- {μ-SBu^t}][c] (not given)	tr P-1 4	13.608)2) 16.945(1) 8.599(1)	98.52(1) 99.00(1) 113.03(1)	FeC$_3$OS	OC 1.80(1,4) μac,O 1.998(5,6) μS 2.245(2,7)	2.589(1) S 70.40(6)	C,C 94.6(4,7.8) C,O 88.6(4,1.8) 176.7(3,1.7) C,S 101.2(3,4.2) 152.3(3,1.0) O,S 83.4(1)	230
				FeC$_3$OS	OC 1.80(1,4) μac,O 1.995(5,4) μS 2.245(2,5)	2.590(1) S 70.48(7)	C,C 94.4(4,7.8) C,O 88.8(4,2.7) 176.7(3,1.6) C,S 101.5(3,5.1) 151.6(3,5) O,S 83.2(2)	
[(OC)$_6$Fe$_2${μ-η^2-(Pr$^i_2$NP)$_2$. NHCCHCO$_2$Et}] (yellow)	m Cc 4	18.211(1) 10.762(2) 16.434(1)	106.09	FeC$_3$P$_2$	OC 1.79(1,5) μP 2.194(3,20) 2.248(3,17)	2.592(2) P 71.4(1,1.0)	C,C 99.0(7,7.9) P,P 73.4(1,7) C,P 98.7(5,5.3) 155.1(4,3.1)	226
[(OC)$_6$Fe$_2${μ-η^2-(Pr$^i_2$NP)$_2$. NCMe}] (red)	m P2$_1$/m 2	8.761(6) 14.362(6) 10.984(1)	100.75(6)	FeC$_3$P$_2$	OC 1.783(7,21) μP 2.228(2,14)	2.593(2) P 71.71(6)	C,C 98.9(3,8.2) P,P 72.72(6) C,P 97.0(2,4.0) 154.5(3,0)	226

Table 5. Continued 16

$[(OC)_6Fe_2\{\mu\text{-tatc}\}.$ $\{\mu\text{-CO}\}]\cdot 0.5C_6H_6$ (blue)	m C2/c 8	23.765(5) 12.402(1) 15.951(3)	 118.17	FeC_4N	OC not given μOC not given N not given	2.593(2) not given	not given	231
$[N(H)Et_3][(OC)_6Fe_2.$ $\{\mu\text{-SO}_2\}\{\mu\text{-SBu}^t\}]$ (red)	m $P2_1$ 2	9.2071(14) 9.3672(18) 14.1126(26)	 93.31(1)	FeC_3S_2	OC 1.789(15,31) μO_2S 2.211(7,3) μBu^tS 2.270(8,4)	2.593(8) S 71.8(2) 69.7(2)	C,C 96.5(7,9.3) S,S 78.0(2,2) C,S 96.3(5,8.3) 158.8(5,1.9)	232
$[(OC)_6Fe_2\{\mu\text{-}(Pr^i{}_2NP).$ $(Bu^t{}_2PP)CO\}]$ (orange)	m $P2_1/c$ 4	8.901(1) 29.590(7) 12.136(4)	 112.97(2)	FeC_3P_2	OC 1.76(2,3) μP 2.236(4,30)	2.593(3) P 70.8(1,8)	P,P 69.5(1,1)	233
$[(OC)_2Fe_2\{\mu\text{-H}\}\{\mu\text{-SMe}\}.$ $(PMe_2Ph)]PF_6$ (not given)	tr P-1 2	13.804(3) 13.834(3) 8.890(2)	105.41(2) 92.22(2) 103.11(2)	FeC_2S_2HP	OC 1.748 μH 1.600 μS 2.274 P 2.242	2.595 H not given S 69.5	S,S 76.7	234
$[(OC)_6Fe_2\{\mu\text{-}\eta^2\text{-}(Pr^i{}_2NP)_2\text{-}$ OCOCOCHCH}] (yellow)	tr P-1 2	9.679(16) 10.476(6) 15.440(3)	75.23(9) 73.73(6) 76.40(9)	FeC_3P_2	OC 1.75(3,11) μP 2.165(9,10) 2.214(8,7)	2.595(6) P 72.6(2,9)	C,C 99(1,7) P,P 73.8(3,3) C,P 97(1,9) 152(1,3)	226
$[(OC)_6Fe_2\{\mu\text{-}C_{31}H_{48}O_2P_2\}]$ $\cdot C_3H_6$ (not given)	m $P2_1/n$ 4	15.251(2) 16.557(3) 17.355(3)	 96.93(2)	FeC_3P_2	OC not given μP 2.224(1,9)	2.5961(7) P 71.41(3,3)	P,P 63.67(4,15)	217
$(PPh_4)[(OC)_6Fe_2.$ $\{\mu\text{-}\eta^2\text{-AsS}\}]$ orange yellow) (at 146 K)	tr P-1 2	11.409(2) 12.273(3) 12.277(3)	108.17(2) 102.05(2) 106.60(2)	FeC_3SAs	OC 1.776(8,18) μS 2.305(2,6) μAs 2.352(1,5)	2.598(1) S 68.63(3) As 67.07(4)	C,C 100.4(3,2.6) S,As 57.15(5,1) C,S 100.3(2,4.1) 148.1(1) C,As 98.3(2,1.7) 150.1(2,0)	235
$[(OC)_6Fe_2\{\mu\text{-P}(Pr^i)SePh\}.$ $\{\mu\text{-SePh}\}]$	tr P-1 2	7.422(3) 14.525(6) 24.047(9)	101.11(3) 94.89(3) 92.99(3)	FeC_3SAs	OC not given μP 2.226(3,7) μSe 2.394(2,2)	2.600(2) P 71.4(1) Se 65.8(1)	not given	222

Table 5. Continued 17

$(NEt_4)[(OC)_6Fe_2\{\mu\text{-}CO\}.$ $\{\mu\text{-}PPh_2\}]$ (red)	tr P-1 2	11.140(1) 13.982(4) 10.161(3)	105.30(2) 103.33(2) 77.2(2)	FeC_4	OC 1.778(3,24) μOC 1.963(3,13) μP 2.210(1,1)	2.601(1) C 82.9(1) P 72.09(3)	not given	236
$[(OC)_6Fe_2\{\mu\text{-}\eta^2\text{-}Pr^i_2NP)_3\}]$ (orange)	m $P2_1/n$ 4	11.554(2) 14.294(6) 20.405(4)	 90.96(2)	FeC_3P_2	OC not given μP 2.248(2)	2.602(2) P not given	P,P 34.4 60.8	227
$[(OC)_6Fe_2\{\mu\text{-}\eta^2\text{-}P_3Bu^t\}]$ (yellow)	or $P2_12_12_1$ 4	10.522(5) 12.188(4) 19.807(8)		FeC_3P_2	OC 1.74(2,3) μP 2.235(5,8)	2.602(3) P 71.1(1,2)	C,C 95.9(10,9.7) P,P 68.0(2,1) C,P 100.7(6,7.9) 154.1(6,3.9)	237
$[(OC)_6Fe_2\{\mu\text{-}Pr^i_2NP)_2CO\}]$ (orange)	m $P2_1/n$ 4	10.197(3) 31.403(13) 9.170(3)	 112.18(2)	FeC_3P_2	OC not given μP not given	2.603(2) P not given	not given	238
$[(OC)_6Fe_2\{\mu\text{-}H\}\{\mu\text{-}CO\}\{\mu\text{-}$ $PBu^t_2\}]$ (not given)	m $P2_1/n$ 4	8.914(9) 14.897(2) 14.632(2)	 99.03(1)	FeC_4HP	OC 1.806(2,27) μOC 1.977(2,21) μH 1.69(2,5) μP 2.2738(6,1)	2.6034(4) H 100.8(1.1) C 82.34(7) P 69.85(2)	C,P 82.39(6,45) 98.40(7,4.52) 160.77(7,3.91)	239
$[(OC)_4Fe_2\{\mu\text{-}CO\}_2\text{-}$ $(bpy)(PBu^n_3)]$ (not given)	tr P-1 2	10.727(1) 14.517(2) 10.306(2)	97.30(1) 96.53(1) 74.13(2)	FeC_5	OC 1.774(8,11) μOC 1.805(7) 2.133(6)	2.6048(14) C 80.3(4,1.1)	C,C 94.1(5,12.4) 159.9(3,1.6)	240
				FeC_3N_2P	OC 1.754(7) μOC 1.847(6) 2.244(6) bpy,N 2.005(5,6) Bu^n_3P 2.279(2)		C,C 93.0(4,4.4) N,N 79.8(3) C,N 87.1(4,8.3) 171.2(3,2.8) C,P 91.6(3,2) 171.5(2) N,P 93.9(2,1.3)	

Table 5. Continued 18

[(OC)$_5$Fe$_2${μ-CO}$_2$(bpy)] (red purple)	not given			FeC$_4$N$_2$	OC not given μOC 2.13(-,24) bpy,N not given	2.611 C not given	not given	241
				FeC$_5$	OC not given μOC 1.94(-,14)		not given	
[(OC)$_6$Fe$_2${μ-(Pr$^i_2$NP)$_2$. CHSiMe$_3$}] (orange)	m P2$_1$ 2	9.771(2) 16.220(1) 10.559(2)	114.34	FeC$_3$P$_2$	OC not given μP 2.231(4,8)	2.612(3) P not given	not given	242
[(OC)$_6$Fe$_2${μ-(Pr$^i_2$NP)$_2$. C(CH$_2$Ph(OH)}] (yellow)	m P2$_1$ 4	9.797(1) 16.593(1) 19.671(2)	101.564(9)	FeC$_3$P$_2$	OC 1.771(9,21) μP 2.231(2,29)	2.613(2) P 71.7(1,4)	C,C 98.6(4,5.8) P,P 66.38(7,45) C,P 99.5(3,8.7) 150.8(3,1.9)	243
[(OC)$_6$Fe$_2${μ-Bu$^t_2$NP}] (orange)	m P2$_1$/n 2	8.198(1) 15.331(2) 15.242(2)	93.05(1)	FeC$_3$NP	OC not given μN 2.006(7,5) μP 2.147(3,1)	2.615(2) N not given P not given	not given	244
[(OC)$_6$Fe$_2${μ-SPh}. {μ-PPh$_2$}]^c (red brown)	or Pna2 8	16.544(1) 20.256(4) 28.76(1)		FeC$_3$SP	OC 1.796(23,77) μS 2.280(5,1) μP 2.238(5,1)	2.614(3) S 69.97(13) P 71.47(26)	S,P 77.87(21,1)	245
				FeC$_3$SP	OC 1.742(28,41) μS 2.261(5,11) μP 2.221(5,6)	2.606(3) S 70.39(17) P 71.86(29)	S,P 77.71(22,36)	
[K.crypt]$_2$[(OC)$_6$Fe$_2$. {μ-η^2-Te$_3$}]^c (dark)	tr P-1 4	11.531(7) 22.098(12) 25.341(13)	85.22(4) 84.60(4) 84.39(5)	FeC$_3$Te$_2$	OC 1.728(28,86) μTe 2.584(4,2) 2.561(3,7)	2.614(4) Te 61.1(1,3)	C,C 99.8(11,4.2) C,Te 94.1(8,5.9) 157.1(7,2.4)	246
				FeC$_3$Te$_2$	OC 1.757(30,52) μTe 2.580(4,1) 2.564(4,5)	2.638(5) Te 61.7(1,2)	C,C 98.2(11,5.6) C,Te 95.3(10,8.9) 157.6(9,3.2)	

Table 5. Continued 19

[(OC)$_6$Fe$_2${μ-PPh$_2$}. {P(OMe)$_3$}] (green) (at 179 K)	m Cc 4	14.629(4) 11.218(4) 15.004(3)	92.60(2)	FeC$_3$P	OC 1.803(4,16) μP 2.205(1)	2.6159(9) P 72.50(3)	C,C 96.8(2,1.3) P,P 99.4(1,5) 155.6(1)	247
				FeC$_3$P$_2$	OC 1.786(4,11) μP 2.219(1) P 2.175(1)		C,C 106.3(2,2.0) 147.3(2) P,P 164.04(4) C,P 91.1(1,7.3)	
[(OC)$_6$Fe$_2${μ-(Pr$^i_2$NP). (ClP)OCPh$_2$}] (orange)	m P2$_1$ 4	15.467(1) 10.589(1) 17.263(2)	96.124(9)	FeC$_3$P$_2$	OC 1.79(2,7) μP 2.192(5,2)	2.616(3) P 73.3(2,2)	C,C 98.6(10,7.4) P,P 71.0(2,1) C,P 98.6(8,7.7) 151.8(7,6.8)	248
[(OC)$_6$Fe$_2${μ-CO}. {μ-η^2-(Me$_4$CNNC}] (dark green)	m P2$_1$/n 4	9.816(1) 14.046(1) 12.936(1)	98.631(6)	FeC$_4$N	OC 1.806(3,13) μOC 2.020(2,49) N 1.907(2,4)	2.623(1) C 80.9(1)	C,C 89.6(1,6.4) 106.5(1,1.8) C,N 88.4(1,4.8)	249
[(OC)$_6$Fe$_2${μ-Te(CH$_2$)$_2$Te}] (orange red)	m P2$_1$/n 4	7.194(1) 13.179(4) 14.644(3)	91.57(2)	FeC$_3$Te$_2$	OC 1.78(1,1) Te 2.562(1,8)	2.626(2) Te 62.65(4,9)	Te,Te 83.29(4,12)	250
ppn[(OC)$_6$Fe$_2${μ-CO}. {μ-PBu$^t_2$} (not given	tr P-1 2	10.709(3) 14.498(3) 17.224(4)	68.84(1) 77.42(1) 84.71(2)	FeC$_4$P	OC 1.771(4,12) μOC 1.964(3,34) μP 2.249(1,1)	2.627(1) C 83.91(13) P 71.49(3)	P,C 81.82(9,75) 99.47(12,6.00) 154.37(12,4.19)	239
[(OC)$_6$Fe$_2${μ-η^2-P(Ph)-1,2- CH$_2$C$_6$H$_4$CH$_2$PPh}] (orange red)	m P2$_1$/c 4	11.996(1) 13.592(2) 16.891(2)	109.80(1)	FeC$_3$P$_2$	OC not given μP 2.222(2,8)	2.628(2) P not given	not given	251
(NEt$_4$)[(OC)$_6$Fe$_2${μ-Te$_2$Br}] (orange red)	m P2$_1$/n 4	13.304(2) 13.022(3) 13.372(5)	93.70(2)	FeC$_3$Te$_2$	OC 1.78(2,5) μTe 2.544(3,23)	2.629(3) Te 62.24(8,48)	Te,Te 64.95(7,6)	250

Table 5. Continued 20

$[(OC)_3Fe_2\{\mu\text{-}Pcy_2\}.$ $\{\mu\text{-}PBu^t{}_2\}\{\mu\text{-}dppm]\cdot CH_2Cl_2$ (violet brown)	m Cc 4	21.930(3) 12.845(1) 18.153(2)	103.345(8)	FeP_3C	μcy_2P 2.184(1) $\mu Bu^t{}_2P$ 2.223(1) μdppm,P 2.229(1) OC 1.737(3)	2.631(1) P 72.06(3,3)	P,P 92.45(3,4.42) 145.60(2) P,C 106.08(9,1.51)	252
				FeP_3C_2	μcy_2P 2.288(1) $\mu Bu^t{}_2P$ 2.248(1) μdppm,P 2.247(1) OC 1.754(3,2)		P,P 89.52(3,4.62) 135.30(3) C,C 93.0(1) P,C 98.7(10,14.81) 173.61(10)	
$[(OC)_6Fe_2\{\mu\text{-}PPh\}_4\}]$ (not given)	m $P2_1/c$ 4	16.988(6) 11.492(2) 16.336(4)	101.09(2)	FeC_3P_2	OC 1.783(4,21) μP 2.220(1,6)	2.633(1) P 72.8(0,1)	C,C 97.5(2,5.0) P,P 74.6(0,1) C,P 98.1(2,12.5) 153.8(2,6.4)	253
$[(OC)_6Fe_2\{\mu\text{-}Te(CH_2)_3Te\}]$ (orange red)	m $P2_1/n$ 4	7.455(1) 13.312(1) 15.097(3)	91.48(2)	FeC_3Te_2	OC 1.784(7,8) μTe 2.5345(10,59)	2.633(1) Te 62.60(3,4)	Te,Te 87.08(3,7)	250
$[(OC)_6Fe_2\{\mu\text{-}PMe\}_4\}]$ (not given)	m $P2_1/c$ 4	7.906(4) 17.139(6) 13.849(6)	101.41(3)	FeC_3P_2	OC 1.795(13,22) μP 2.219(3,4)	2.634(2) P 72.9(1,1)	C,C 97.9(6,5.0) P,P 75.5(1,1) C,P 97.3(4,8.9) 154.5(4,2.8)	253
$[(OC)_6Fe_2\{\mu\text{-}TeMe\}_2]$ (red) (at 193 K)	or Pmna 4	20.902(5) 10.840(19) 6.306(8)		FeC_3Te_2	OC 1.78(1,0) μTe 2.550(3,2)	2.634(5) Te 62.3(1,1)	C,C 97.9(5,5.1) Te,Te 79.52(7) C,Te 94.7(4,5.3) 159.4(4,1.4)	254
$[(OC)_6Fe_2\{\mu\text{-}TeCHCl_2)_2\}]$ (orange red)	m Cc 4	13.443(3) 14.971(3) 9.271(1)	90.25(1)	FeC_3Te_2	OC 1.79(3,7) μTe 2.530(4,20)	2.656(5) Te 63.34(12,29)	Te,Te 83.88(11,25)	224
$[(OC)_6Fe_2\{\mu\text{-}Bu^tPS_2PBu^t\}]$ (orange)	or $Pna2_1$ 4	24.331(3) 9.0675(8) 9.7343(7)		FeC_3P_2	OC 1.795(12,34) μP 2.213(3,11)	2.659(2) P 73.9(1,2)	P,P 76.1(1,1) C,P 98.2(4,14.8) 156.3(4,14.0)	255

Table 5. Continued 21

$(OC)_6Fe_2(\mu\text{-}CO)$ $\{\mu\text{-}\eta^2\text{-}MeN(P(OMe)_2)_2\}$ (orange)	m $P2_1$ 2	9.1429(17) 15.9765(51) 6.8758(13)	 101.00(1)	FeC_4P	OC 1.792(2,37) μOC 1.972(2) μP 2.1919(5)	2.6606(5) C 84.9(1)	C,C 97.2(1,16.0) 143.1(1) C,P 89.5(1,2.6) 174.0(1)	256
$[(OC)_5Fe_2\{\mu\text{-}CO\}\text{-}$ $\{\mu\text{-}(PF_2)_2NMe\}_2]$ (orange)	or Pccn 8	10.061 9.269 19.876		FeC_3P_2	OC not given μOC not given P not given	2.661(1) C not given	P,P 172.5(1)	257
$[(OC)_6Fe_2\{\mu\text{-}P(H)(CF_3)\}_2]$ (lemon)	tr P-1 2	8.112(2) 8.287(3) 12.891(2)	79.56(1) 80.88(1) 69.52(2)	FeC_3P_2	OC 1.790(11,19) μP 2.193(3,4)	2.661(2) P 74.7(1,1)	C,C 97.2(5,7.5) P,P 79.0(1,1) C,P 96.9(4,10.0) 155.1(4,1.9)	258
$[(OC)_6Fe_2\{\mu\text{-}P(H)(CF_3)\}_2]$ (lemon)	m $P2_1/c$ 4	17.20(1) 11.41(1) 14.91(2)	 92.57(7)	FeC_3P_2	not given	not given	not given	258
$[(OC)_4Fe_2\{\mu\text{-}CO\}.$ $\{\mu\text{-}(EtO)_2POP(OEt)_2\}_2]^c$ (not given)	tr P-1 4	17.850(5) 18.082(6) 11.270(2)	97.26(2) 97.12(2) 102.69(2)	FeC_3P_2	OC 1.697(23,20) μOC 1.916((15,4) P 2.149(4,16)	2.663(2) C 88.03(65)	C,C 107.7(1,1.0) 126.1(1,20.3) P,P 173.3(2,2)	259
				FeC_3P_2	OC 1.713(14,28) μOC 1.938(11,10) P 2.154(5,6)	2.669(2) C 87.07(44)	C,C 108.5(6,4) 125.7(5,19.0) P,P 173.9(1,1.8)	
$[(OC)_3Fe_2\{\mu\text{-}PPhCl\}_2]$ (orange yellow)	tr P-1 4	12.837 19.145 11.416	114.16 100.19 92.74	FeC_3P_2	OC 1.78(2,4) μP 2.20(0,0)	2.67(0) P 74.7(1,0)	P,P 76.7(1,2.0)	210
$[(OC)_6Fe_2\{\mu\text{-}Pr^i_2NPH\}.$ $\{\mu\text{-}norC_7H_8P\}]$ (yellow)	tr P-1 2	7.783(7) 9.124(4) 17.635(8)	98.67(4) 94.23(5) 104.91(5)	FeC_3P_2	OC 1.77(1,1) μP 2.181(3,2) 2.206(2,4)	2.674(2) P 75.12(8,50)	C,C 100.0(5,8.7) P,P 77.00(8,12) C,P 94.5(4,8.1) 155.8(4,2.2)	226

Table 5. Continued 22

[(OC)$_6$Fe$_2${μ-P(Ph)OCH$_2$. C$_6$H$_4$CHP(Ph)(Me)}] (orange)	m P2$_1$/n 4	11.670(2) 14.581(2) 16.674(2)	109.47(1)	FeC$_3$P$_2$	OC not given μP 2.228(1,7)	2.684(1) P not given	not given	251
				FeC$_4$P$_2$	OC not given LC 2.139(3) μP 2.190(1,0)		not given	
[(OC)$_4$Fe$_2${μ-PPh$_2$}. (PMe$_3$)$_2$(NO)] (dark red)	m P2$_1$/a 4	15.509(5) 9.781(2) 19.159(8)	109.46(3)	FeC$_3$P$_2$	OC 1.76(3,2) μP 2.225(6) Me$_3$P 2.231(7)	2.684(4) P 74.2(2)	C,C 95(1,1) 169.1(9) P,P 155.4(3) C,P 93.2(1,9.8)	
				FeP$_2$NC	μP 2.224(7) Me$_3$P 2.193(7) ON 1.69(2) OC 1.63(3)		P,P 100.2(3) P,N 107.4(7,11.9) P,C 88.3(1) 123(1) N,C 115(1)	260
[(OC)$_5$Fe$_2${μ-Pcy$_2$}. {μ-PBu$^t_2$}(PHPh$_2$)]^c (red brown)	m P2$_1$/n 8	17.184(4) 26.641(3) 17.756(4)	105.88(1)	FeC$_3$P$_2$	OC 1.752(27,55) μP 2.263(3,12)	2.686(2) P 73.47(9,37)	P,P 105.27(9) C,P 96.4(6,18.0) 159.4(9,5.0)	252
				FeP$_3$C$_2$	P 2.189(2) μP 2.228(2,8) OC 1.782(9,6)		P,P 105.67(9) 107.70(9) 146.49(8) C,P 95.9(3,11.2)	
				FeC$_3$P$_2$	OC 1.765(11,28) μP 2.267(3,10)	2.717(2) P 74.42(9,23)	P,P 104.22(9) C,P 97.9(4,16.1) 163.6(3)	
				FeP$_3$C$_2$	P 2.197(3) μP 2.226(3,2) OC 1.770(9,2)		P,P 105.29(10) 106.95(9) 147.75(9) C,P 97.0(3,11.5)	

Table 5. Continued 23

[$(OC)_6Fe_2${μ- $C_{33}H_{68}B_4N_4P_2$}]·tol (brown)	m Cc 4	13.368(14) 19.17(2) 19.54(2)	 92.50(9)	FeC_3P	OC not given μP 2.240(1)	2.692(1) P not given	C,P 105.5(2) 139.8(2)	261
				FeC_3P_2	OC not given P 2.262(1) μP 2.289(1)		C,C 171.1(3)	
[$(OC)_6Fe_2${μ-PPh_2}(NO)] (dark red)	tr P-1 2	8.776(1) 9.1907(1) 13.095(1)	84.356(7) 105.65(1) 95.63(1)	FeC_4P	OC 1.810(5,20) μP 2.260(1)	2.699(1) P 74.62(4)	C,C 92.7(2,4.9) 103.0(2) 166.1(2) C,P 92.5(1,7.4) 157.0(2)	260
				FeC_2NP	OC 1.776(6,23) ON 1.688(5) μP 2.192(1)		C,C 99.4(2,3.7) C,P 100.2(2) 135.8(2) C,N 109.0(2,4.8) N,P 106.5(2)	
[$(OC)_5Fe_2${μ-PPh_2}. {μ-$PhPCH_2PPh_2$}] (orange)	m $P2_1/c$ 4	14.8348(53) 9.7766(28) 23.3866(83)	 90.349(29)	FeC_3P_2	OC 1.787(13,13) μPPh_2 2.241(3,46) μP 2.226(3,1) P 2.224(3)	2.699(2) P 74.3(1,3)	P,P 80.8(1)	207
				FeP_3C_2	OC 1.757(12,13) μP 2.215(3,20)		P,P 78.2(1,4.7) 149.9(3)	
[$(OC)_6Fe_2${μ-PPh_2}. {μ-P(Bu^t)$_2$}$_2$] (orange)	m $P2_1/c$ 4	10.116(3) 16.810(4) 16.841(4)	 105.38(2)	FeC_3P_2	OC not given μPh_2P 2.246(1) μBu^t_2P 2.269(1)	2.7017(9) P not given	not given	167
[$(OC)_6Fe_2${μ-$SiMe_2$}$_2$. {μ-SiMeCl}] (red)	hx $P6_3/m$ 2	9.5660(10) 12.042(2)		FeC_3Si_3	OC 1.777(6) μSi 2.322(2)	2.705(1) Si 71.2(1)	C,C 97.8(3) Si,Si 89.5(1) C,Si 86.2(3,3) 173.9(2)	262

Table 5. Continued 24

$[(OC)_6Fe_2\{\mu\text{-}PPh_2\}.$ $\{\mu\text{-}CH_2C(Ph)NMe\}]$ (yellow)	m $P2_1/c$ 4	9.067(2) 23.178(4) 16.527(5)	90.53(2)	FeC_4P	OC 1.790(7,11) µLC 2.106(7) µP 2.194(2)	2.707(1) P 75.6(0)	C,C 92.9(2,13.7) 178.0(2) C,P 95.6(2,9.7) 147.2(2)	
				FeC_3NP	OC 1.795(7,28) µLN 2.029(5) µP 2.222(2)		C,C 98.2(2,8.6) C,N 90.9(2,2.4) 175.4(2) C,P 96.9(2,6.1) 150.2(2) N,P 87.8(1)	263
$[(OC)_6Fe_2\{\mu\text{-}CO\}.$ $\{\mu\text{-}tppm\}]$ (brown)	m $P2_1/c$ 4	11.257(2) 11.926(3) 22.911(8)	100.22(2)	FeC_4P	µOC 2.00(1,2) OC 1.79(1,5) P 2.254(3,4)	2.709(2) C 85.6(5)	C,C 87.8(4) 96.3(6,14.9) 148.7(6) C,P 89.2(4,7.7) 92.0(3,1) 171.5(4,4.2)	264
$[(OC)_5Fe_2\{\mu\text{-}Pcy_2\}.$ $\{\mu\text{-}PBu^t_2\}(PBu^n_3)]$ (red brown)	m Cc 4	25.772(5) 11.891(3) 19.583(4)	135.385(8)	FeC_3P_2	OC 1.775(13,20) µP 2.273(3,1)	2.709(1) P 73.85(12,15)	P,P 104.89(11) C,P 97.8(5,12.8) 160.4(4)	252
				FeP_3C_2	P 2.246(2) µP 2.238(4,9) OC 1.769(10,1)		P,P 107.26(12) 115.75(12) 137.00(11) P,C 95.2(4,8.8)	
$[(OC)_4Fe_2\{\mu\text{-}CO\}.$ $\{\mu\text{-}dppm\}_2]\cdot 2tol$ (brown)	or Pnnm 4	23.409(3) 18.524(3) 14.219(9)		FeC_3P_2	µOC 1.959(6) OC 1.758(6,31) µdppm,P 2.212(2,0)	2.711(1) C not given	C,C 138.8(3) P,P 170.35(7) C,P 88.9(2)	265
$[(OC)_4Fe_2\{\mu\text{-}PPh_2\}_2.$ $\{\mu\text{-}dppm\}]$ (bright orange)	m $P2_1/m$ 4	13.6589(53) 18.1858(50) 18.3919(47)	91.400(27)	FeP_3C_2	μPh_2P 2.235(2,17) µdppm,P 2.247(2,24) OC 1.762(7,14)	2.713(2) P 74.7(1,4)	P,P 87.3(1,9.3) 141.2(1,5.0)	207

Table 5. Continued 25

Compound	Cryst. cl. Space group Z	a b c	β	Chromophore	Fe–L	Fe–Fe Fe–X–Fe	L–Fe–L	Ref.
$[(OC)_5Fe_2\{\mu\text{-}P(2\text{-furyl})_2\}_2$. $\{P(2\text{-furyl})_3\}]$ (not given)	or $Pna2_1$ 4	22.553(6) 16.552(6) 9.250(1)		FeC_3P_2	OC 1.75(1,1) μP 2.245(3,4)	2.715(2) P 75.43(8,3)	C,C 98.4(5,9.5) P,P 78.6(1) C,P 96.4(4,9.3) 155.1(4,9.1)	266
				FeP_3C_2	P 2.203(2) μP 2.194(2,3) OC 1.79(1,3)		P,P 87.95(9,7.12) 171.7(1) C,C 104.9(5) C,P 94.4(3,9.4) 150.7(4)	
$[(OC)_8Fe_2\{\mu\text{-}SO_2\}]$	m $P2_1/c$ 4	8.75(4) 13.19(2) 12.74(2)	113.4(2)	FeC_4S	OC 1.790(22,27) μS 2.216(5,4)	2.717(4) S 75.6(2)	C,C 93.1(8,16.6) 176.5(8,1.5) C,S 94.2(6,8.2) 150.3(6,2.4)	267
$[(OC)_7Fe_2\{\mu\text{-}P(Bu^t)$. $N(Bu^t)CO\}]$ (red)	m $P2_1/c$ 12	19.839(10) 12.787(7) 26.578(10)	110.49(3)	FeC_4P	OC not given μP 2.203(3,11)	2.718(2) P 75.05(9)	not given	244
$[(OC)_4Fe_2\{\mu\text{-}CO\}$. $\{\mu\text{-}dmpm\}_2]$ (orange)	or Pnma 4	9.789(1) 16.656(3) 14.448(2)		FeC_3P_2	OC 1.757(9,25) μOC 1.931(7) P 2.201(4)	2.719(4) C 89.5(4)	C,C 110.2(4,6.8) 139.6(3) P,P 171.4(1) C,P 89.6(3,3.4)	268
$[(OC)_5Fe_2\{\mu\text{-}PBu^t_2\}$. $\{\mu\text{-}Pcy_2\}\{\eta^1\text{-}dppm\}]$ (violet brown)	m $P2_1/n$ 4	11.803(2) 22.930(4) 22.447(4)	104.77(1)	FeC_3P_2	OC 1.769(5,14) μP 2.278(1,10)	2.719(1) P 74.16(4,21)	C,C 86.6(2,2.2) 133.4(2) P,P 104.23(4)	269
				FeP_3C_2	P 2.242(1) μP 2.231(1,1) OC 1.776(4,3)		P,P 107.45(4) 126.28(4,12.71) C,C 156.8(2) P,C 84.9(1,3) 94.3(1,8.9)	

Table 5. Continued 26

Compound	Crystal data	a, b, c	β	Chromophore	Fe–L	Fe–Fe / Fe–X–Fe	L–Fe–L	Ref.
[(OC)$_8$Fe$_2${μ-SiMes$_2$}]^c (orange)	tr P-1 4	14.342(3) 17.508(4) 12.617(2)	106.54(2) 90.77(2) 112.52(2)	FeC$_4$Si	OC 1.806(11,30) μSi 2.405(2,8)	2.725(2) Si 69.10(7)	not given	270
				FeC$_4$Si	OC 1.816(11,32) μSi 2.402(2,11)		not given	
[Fe$_2${μ-(PF$_2$)$_2$NMe}$_3$. {(PF$_2$)$_2$NMe}(CO)] (yellow)	m C2/c 8	31.023(7) 9.273(3) 17.401(5)	96.57	FeP$_5$	P not given μP not given	2.725(2) P not given	not given	271
				FeP$_4$C	P not given μP not given OC not given		not given	
[(OC)$_4$Fe$_2${μ-Br}{μ-PhO)$_2$ PN(Et) P(OPh)$_2$}$_2$]PF$_6$· Me$_2$CO (not given)	or Cmc2$_1$ 4	16.82(2) 19.30(2) 20.82(2)		FeC$_2$P$_2$Br	OC 1.78(3,9) P 2.221(5,14) μBr 2.377(5,14)	2.725(5) Br 69.9(1)	not given	272
[(OC)$_6$Fe$_2${μ-PPh$_2$}. {μ-η^2-Ph$_2$PC(CN)PPh$_2$}] (yellow)	m P2$_1$/c 4	13.033(3) 13.124(6) 23.929(5)	96.89(2)	FeC$_3$P$_2$	OC 1.790(4,29) μPPh$_2$ 2.211(1,13) P 2.290(1,11)	2.727(1) P 76.10(3)	C,C 89.4(2,2.7) 175.9(2,5) P,P 146.3(1,2.2) C,P 95.7(1,13.1)	273
[(OC)$_6$Fe$_2${μ-Bu^{t}As. NSNAsBu^t}] (red)	or Pc2$_1$n 4	8.496(3) 14.791(2) 17.669(3)		FeC$_3$As$_2$	OC 1.778(5,9) μAs 2.3241(7,7)	2.7285(9) As 71.89(2)	C,C 96.5(2,8.7) As,As 79.60(3,3) C,As 94.9(1,6.5)	274
[(OC)$_6$Fe$_2${μ-P$_2$Bu$^t_2$}] (red)	m P2$_1$/n 4	15.435(2) 15.431(2) 8.539(1)	91.34(1)	FeC$_3$P$_2$	OC not given μP 2.222(2,10)	2.740(1) P not given	not given	275

Table 5. Continued 27

[(OC)$_4$Fe$_2${μ-η^2-S(CH). Nallyl}{μ-Pcy$_2$}{μ-dppm}] (red)	tr P-1 2	11.9777(39) 13.5843(36) 13.7530(51)	88.912(26) 89.293(28) 78.931(25)	FeC$_2$P$_2$N	OC 1.756(8,19) μP 2.247(2) μdppm,P 2.277(2) μL,N 2.031(6)	2.741(2) P 75.4(1)	P,P 146.6(1) P,N 90.8(1)	
				FeC$_2$P$_2$S	OC 1.747(8,20) μP 2.234(2) μdppm,P 2.245(2) μL,S 2.324(2)		P,P 144.9(1) P,S 92.0(1)	221
[(OC)$_4$Fe$_2${μ-CO}. {μ-dppm}$_2$] (dark brown)	tr P-1 2	13.549(8) 14.455(9) 16.065(9)	89.99(4) 68.75(6) 68.63(5)	FeC$_3$P$_2$	OC 1.80(2,5) μOC 1.99(2,1) P 2.265(7,10)	2.742(4) C 86.9(8)	C,C 98.1(10,2.0) 160.3(9) P,P 172.9(3,1.5) C,P 89.8(6,4.3)	276
[(OC)$_6$Fe$_2${μ-(Pr$^i_2$NP)$_2$H. COCHP(C$_6$H$_4$Me-p}] (yellow)	m P2$_1$ 2	10.925(6) 14.822(4) 14.581(6)	109.05(3)	FeC$_3$P$_2$	OC 1.80(2,7) μP 2.249(4,59)	2.747(2) P 75.1(1)	P,P 74.9(2,1.7)	277
[(OC)$_8$Fe$_2${μ-SiPh$_2$H}$_2$] (yellow)	tr P-1 2	10.018(2) 10.436(2) 14.881(3)	108.00(3) 91.46(3) 96.97(3)	FeC$_2$Si$_2$H	OC 1.80(2,7) μSi 2.382(6,24) H 1.57	2.759(6) Si 70.8(2)	C,C 86.3(10,1.6) 170.1(9) Si,Si 109.2(2) C,Si 87.7(7,6) 103.8(8)	278
[(OC)$_7$Fe$_2${μ-η^2-SiPh$_2$H}. (SiPh$_2$H)] (red)	m C2/c 8	29.079(6) 9.731(2) 21.113(3)	97.10(3)	FeC$_4$Si	OC 1.803(8,16) μSi 2.301(2)	2.769(1) Si 72.1(1)	C,C 97.1(3,5.9) 170.0(3) C,Si 90.2(2,7.7) 159.1(2)	278
				FeC$_3$Si$_2$H	OC 1.792(8,7) μSi 2.385(2) Si 2.365(2) H 1.66		C,C 92.1(3,9) 161.8(3) Si,Si 131.0(1) C,Si 89.4(2,9.1) 133.4(2)	

Table 5. Continued 28

$[(OC)_5Fe_2\{\mu\text{-}PPh_2\}.$ $(PMe_3)_3][(OC)_8Fe_2.$ $\{\mu\text{-}PPh_2\}]$ (orange red) (at 203 K)	m $P2_1$ 2	10.097(2) 17.077(6) 15.418(7)	 93.60(3)	FeC_3P_2	OC 1.781(8,13) μP 2.230(2) Me_3P 2.230(2)	2.780(2) P 77.37(8)	C,C 91.1(4,2) 176.1(4) P,P 110.5(1) C,P 91.4(3,9.3) 148.8(3)	247
				FeC_2P_3	OC 1.764(8,0) μP 2.217(2) Me_3P 2.245(2,20)		C,C 171.4(4) P,P 104.2(1,5) 151.1(1) C,P 89.9(3,7.8)	
				FeC_4P (2x)	OC 1.762(8,24) μP 2.324(2,8)	not given P 118.7(1)	C,C 90.7(4,2.9) 120.0(4,5.7) C,P 89.3(3,2.5) 177.6(3,2)	
$(N\text{-}metlut)_2[Fe_2(OC)_8]$ (deeply colored)	m $P2_1/n$ 2	10.112(4) 9.410(3) 13.370(5)	 96.84(2)	FeC_4	OC 1.769(3,34)	2.780(1)	C,C 96.6(1,2.2) 118.7(1,3)	279
$(ppn)_2[Fe_2(CO)_6]\cdot 2MeCN$ (red brown)	tr P-1 2	17.942(4) 12.944(3) 10.532(3)	93.21(2) 115.35(1) 106.72(2)	FeC_3	OC 1.743(16) 1.769(17)	2.787(2)	C,C 96.7(6) 118.2(8,5)	280
$[(OC)_5IFe_2\{\mu\text{-}(MeO)_2PN.$ $(Et)P(OMe)_2\}_2]PF_6$ (not given)	or Pbca 8	20.233(9) 15.545(7) 22.218(9)		FeC_3P_2	OC 1.77(2,3) P 2.201(4,5)	2.787(3)	not given	281
				FeC_2P_2I	OC 1.78(2,1) P 2.221(5,4) I 2.667(2)		not given	
$(PPh_4)_2[Fe_2(CO)_8]\cdot 2MeCN$ (not given)	tr P-1 1	9.389(3) 12.086(6) 12.931(11)	71.37(4) 76.75(4) 69.83(4)	FeC_4	OC 1.765(3,29)	2.792(1)	C,C 96.1(1,3.1) 118.9(1,1.3)	282

Table 5. Continued 29

Compound	Cryst. cl. Space Group Z	a [Å] b [Å] c [Å]	β [°]	Chromophore	Fe-L [Å]	Fe-Fe [Å] Fe-L-Fe [°]	L-Fe-L [°]	Ref.
$[(OC)_8Fe_2\{\mu\text{-}SbCH\text{-}(SiMe_3)_2\}]$ (not given)	m $P2_1/n$ 4	19.958(4) 6.635(1) 24.408(6)	103.51(2)	FeC_4Sb	OC not given μSb 2.652(1,9)	2.801(1) Sb 64.15(2)	not given	283
$[(OC)_4Fe_2\{\mu\text{-}PPh_2\}.(PMe_3)_2(NO)]$ (not given)	m $P2_1/a$ 4	12.299(1) 15.649(2) 14.681(1)	95.106(6)	FeC_3P_2	OC 1.772(6,16) μP 2.221(2) Me_3P 2.216(2)	2.812(1) P 79.37(5)	C,C 112.1(3,5.5) 135.6(2) C,P 88.9(2,3.1) P,P 173.27(6)	
				FeP_2NC	μP 2.182(2) Me_3P 2.250(2) ON 1.660(5) OC 1.725(6)		P,P 153.32(6) P,N 100.1(2,3.9) P,C 93.3(2,2.2) N,C 116.3(2)	260
$[(OC)_6Fe_2\{\mu\text{-}P(CF_3)_2\}_2]$ (red)	or Pnma 4	20.26(2) 12.591(2) 7.422(1)		FeC_3P_2	OC 1.814(4,3) μP 2.194(1,1)	2.819(1) P 80.0(1,1)	C,C 94.4(2,8.3) P,P 83.5(1) C,P 96.8(1,6.5) 159.3(1,1.8)	284
$[(OC)_8Fe_2\{\eta{:}\eta\text{-}Ph_2SiOSiPh_2\}]$ (red)	m $P2_1/c$ 4	17.802(4) 17.099(3) 10.919(2)	104.50(3)	FeC_4Si	OC 1.796(11,26) Si 2.419(2,7)	2.875(2) -	C,C 94.5(4,3.6) 166.1(4,9) C,Si 84.0(3,4.5) 173.8(3)	278
$[(OC)_6Fe_2\{\mu\text{-}SMe\}_3].[Fe_2\{\mu\text{-}S_2C_2(CF_3)_2\}_2.\{S_2C_2(CF_3)_2\}]$ (black)	tr P-1 2	13.64(1) 14.81(1) 17.81(1)	124.27(3) 127.19(3) 76.47(3)	FeC_3S_3	OC 1.80(1,3) μS 2.305(4,10)	3.062(4) S 83.2(1,1)	C,C 93.5(5,9) S,S 80.7(1,8) C,S 92.6(4,3.0) 170.8(4,1.6)	
				FeS_5	S 2.187(4,23) μS 2.258(3,58)	2.767(4,11) S 75.6(1,4)	S,S 87.3(1,1.3) 88.5(1,3) 101.4(1,5.2) 104.5(1,5) 155.1(1,5) 156.3(1,3)	285

Table 5. Continued 30

$[(OC)_4Fe_2\{\mu\text{-}SO_2\}.$ $\{\mu\text{-}SMe\}_2(PMe_3)_2].$ $0.5Et_2O^c$ (not given)	trg $I4_1md$ 8	13.164(1) 28.840(2)		FeS_3C_2P	OC 1.749(8,0) μO_2S 2.217(3) μMeS 2.360(2,0) P 2.342(3)	3.115(2) S 89.3(0) 82.6(0)	C,C 94.6(3) S,S 77.9(0,3.7) C,S 94.3(2,1.1) 169.9(2) C,P 91.1(2) S,P 93.2(1) 173.5(0)	286
				FeS_3C_2P	OC 1.745(7,0) μO_2S 2.235(2) μMeS 2.352(2,0) P 2.320(3)	3.119(2) S 88.5(0) 83.1(0)	C,C 94.5(3) S,S 78.1(1,3.3) C,S 93.7(2,9) 168.6(2) C,P 92.0(2) S,P 93.0(0) 172.9(0)	
$[(OC)_6Fe_2\{\mu\text{-}S_2C_2.$ $(CF_3)_2\}_2]$ (dark purple)	m $P2_1/n$ 4	14.455(38) 11.180(36) 7.093(37)	90.30(5)	FeC_3S_3	OC 1.846(20,14) μS 2.303(10,25) S 2.288(11)	3.363 S 95.1(3)	C,C 93.5(9,3.1) S,S 86.3(3,1.4) C,S 90.0(8,5.6) 173.7(8,3.9)	287
$[(OC)Fe\{\mu\text{-}edtbt\}]_2$ (brown red) (at 200 K)	m $P2_1/c$ 4	12.728(5) 15.341(23) 16.336(16)	104.76(8)	FeS_5C	μS 2.313(3,17) S 2.288(3,37) OC 1.744(11,1)	3.420(1) S 95.2(1,1)	S,S 84.3(1,1) 175.7(4,1.4)	288
$[(OC)_6Fe\{\mu\text{-}PMe_2\}_2I_2]$ (red)	m $P2_1/a$ 2	14.633(5) 9.105(3) 8.318(2)	118.41(1)	FeC_3P_2I	OC 1.81(2,7) μP 2.302(6,10) I 2.665(3)	3.590(4) P 102.4(2)	C,C 96.4(9,6.4) P,P 77.6(2) C,P 90.1(7,8) 166.7(7,2) C,I 83.7(7,7) 174.9(7) P,I 93.4(2,2)	289
$(PPh_4)_2[(OC)_4Fe_2.$ $\{\mu\text{-}Se_5P\}_2]$ (brown)	tr P-1 1	10.207(2) 11.151(2) 13.271(2)	81.46(1) 76.03(2) 79.78(2)	$FeSe_4C_2$	μSe 2.468(4,9) Se 2.440(3,32) OC not given	< 3.6 Se 94.0(1)	Se,Se 86.1(1,1) 175.4(2)	290

Table 5. Continued 31

[Na(2,2,2-crypt)]$_2$.	tr	12.466(6)	127.92(2)	FeC_3P_2	OC 1.750(10,21)	3.630(3)	C,C	103.0(4,10.5)	291
[(OC)$_6$Fe{μ-PPh$_2$}$_2$]	P-1	16.861(6)	84.04(2)		μP 2.280(3,9)	P 105.5(1)	P,P	74.5(1)	
(deep burgundy red)	1	10.797(4)	101.83(3)				C,P	94.9(3,10.0)	
								149.7(3,10.2)	
(PPh$_4$)$_2$[(OC)$_4$Fe$_2$.	m	15.552(5)		$FeSe_4C_2$	μSe 2.435(3,2)	3.633(2)	Se,Se	91.3(5,9.1)	292
{μ-Se$_4$Sb}$_2$]	P2$_1$/n	10.279(3)	101.72(3)		Se 2.453(3,2)	Se 96.5(1)		171.2(1)	
(red orange)	2	18.159(6)			OC 1.771(18,4)		C,C	92.8(8)	
							Se,C	87.9(6,2.9)	
								170.0(6,1.4)	
[(OC)$_3$Fe{μ-(PF$_2$)$_2$NMe}]$_2$	m	10.473		FeC_3P_2	OC not given	3.90	C_{eq},C_{eq}	88.1(5,7)	257
(yellow)	P2$_1$/n	10.825	100.127		P not given		C_{eq},P_{eq}		
(at 253 K)	4	9.035					C_{ax},C_{eq}	100.6(7,3.0)	
							C_{ax},P_{eq}		
							P_{eq},P_{ax}	153.9(2)	
(NBu$^n_4$)$_2$[(OC)$_2$Fe.	m	11.268(4)	100.32(3)	$FeTe_4C_2$	μTe 2.616(8,1)	3.921(8)	Te,Te	83.7(3,8)	145
{μ-Te$_4$As}]$_2$	C2/c	21.228(9)			Te 2.616(8,10)	Te 97.1(3)		99.5(3,3)	
(brown)	4	26.413(9)			OC 1.714(63,16)			175.1(4)	
							C,C	97.0(28)	
							Te,C	89.2(18,4.2)	
								168.0(19,9)	
(NEt$_4$)[(OC)$_6$Fe$_2$.	m	11.606(2)		FeC_3P_2	OC 1.763(12,30)	3.961(1)	C,C	120.0(5,10.2)	247
{μ-PPh$_2$}(PPh$_3$)$_2$]	P2$_1$/c	21.647(5)	96.53(2)		μP 2.303(3,12)	P 118.6(1)	P,P	173.1(1,2.9)	
(yellow)	4	21.999(8)			Ph$_3$P 2.195(3,3)		C,P	90.1(3,4.5)	
[(OC)$_8$Fe$_2${μ-P(H)(OMe)}]	m	19.176(2)		FeC_4P	OC 1.76(1,4)	3.995(2)	C,C	92.3(5,2.5)	293
(yellow)	C2/c	16.564(2)	105.08(1)		μP 2.261(3,2)	P 124.1(1)		119.8(5,2.2)	
	8	21.962(2)					C,P	87.6(4,3.9)	
								176.0(5,2.9)	
[(OC)$_8$Fe$_2${μ-AsMe$_2$}]	m	14.006(2)		FeC_4As	OC 1.780(4,11)	not given	C,C	92.6(2,2.6)	294
(orange yellow)	C2/c	16.983(1)	97.16(1)		μAs 2.395(1)	As 120.4(2)		119.8(2,4.6)	
	4	19.041(1)					C,As	87.5(1,7)	
								177.1(1)	

Table 5. Continued 32

Compound	Crystal class / Space group / Z	a, b, c	α, β, γ	Chromophore	Fe–L	Fe–L–Fe	L–Fe–L	Ref.
$[(OC)_6Fe_2\{\eta^2$-μ-μ-$Te_4\}$. {μ-$TeCl_2\}]^c$ (orange)	tr P-1 2	9.045(2) 13.684(3) 17.961(4)	72.74(3) 82.19(3) 75.15(3)	FeC_3Te_3	OC not given μTe 2.624(2)	not given Te 81.94(7)	not given	295
				FeC_3Te_3	OC not given μTe 2.618(2)	not given Te 81.46(6)	not given	
$[(OC)_6Fe_2\{$μ-Cl}. {μ-$TeCl)_2\}]_2[Te_2Cl_{10}]$ (orange)	m $P2_1/c$ 4	16.401(3) 10.594(2) 14.035(3)	115.13(2)	FeC_3Te_2	OC not given μCl 2.334(2) μTe 2.604(2)	not given Cl 86.41(6) Te 75.70(4)	Te,Te 82.26(4)	295
$[(OC)_3Fe\{$μ-POC_6H_2-2,6- Bu^t_2-4-Me$\}]_2$ (red brown)				FeC_3P_2	OC not given μP 2.112(1) 2.202(1)	not given P 113.5(1)	P,P 66.5(1)	296
$[\{(OC)_4Fe\}_2\{$μ-η^2-PMes$\}_2]$ (dark red) (at 130 K)	tr P-1 2	7.980(2) 10.882(3) 17.125(4)	78.49(2) 86.14(2) 82.74(2)	FeC_4P	OC 1.803 P 2.189(1,3)	not given	not given	296
$[\{(OC)_4Fe\}_2$. {μ-η^2-$PN(SiMe_3)_2\}_2]$ (orange red)				FeC_4P	OC 1.794 P 2.232(1)	not given -	not given	296
$[\{(OC)_4Fe\}_2$. {μ-POC_6H_2-2,4,6-$Bu^t_3\}_2]$ (orange red)				FeC_4P	OC not given μP 2.226(2)	not given	not given	296
				FeC_4P_2	OC not given P 2.362(2) μP 2.347(2)	not given	P,P 55.3(1)	
$[\{(OC)_2\{P(OMe)_3\}_2Fe\}_2$- (μ-$N_2$)] (not given)	m $P2_1/c$ 2	12.62(1) 10.876(6) 13.32(1)	120.89(5)	FeC_2P_2N	OC 1.79(1,3) P 2.155(2,3) μN 1.876(9)		C,C 123.4(5) P,P 175.5(1) C,N 118.4(4,5) P,N 88.0(4,2.0)	79

Table 5. Continued 33

Compound	Cryst. cl. / Space group / Z	a, b, c [Å]	α, β, γ [°]	Chromophore	Fe-L [Å]	Fe-X-Fe [°]	L-Fe-L [°]	Ref.
$[(OC)_8Fe_2\{\mu-\eta^2-P(Bu^t)(Cl)-CH_2P(Bu^t)(Cl)\}]$ (yellow)	m C2/c 4	11.181(2) 17.072(2) 13.773(1)	112.46(1)	FeC_4P	OC 1.817(2,33) P 2.222(1)	not given	C,C 89.8(1,3.1) 119.4(1) 174.0(1) C,P 90.6(1,6) 120.3(1,3.9)	217
$[(OC)_8Fe_2\{\mu-P_2(Bu^tN)_2-(H)_2\}]$ (yellow)	not given	not given		FeC_4P	OC 1.794(3,7) P 2.178(1)	not given	C,C 92.7(1,2.6) 112.3(1) 168.7(1) C,P 85.6(1,6) 115.6(1) 132.1(1)	297
$[(OC)_3Fe\{\mu-P(4-Me-2,6-Bu^t_2C_6H_2O\}]_2$ (red brown)	m $P2_1/n$ 2	15.840(2) 8.569(1) 15.996(2)	119.60(1)	FeC_3P_2	OC 1.809(5,16) µP 2.112(1) 2.202(1)	not given P 113.5(1)	C,C 98.0(2,8.0)	298
$[(OC)_8Fe_2\{\mu-P(2,4,6-Bu^t_3C_6H_2O)\}_2]$ (orange red) (at 140 K)	m $P2_1/c$ 4	11.243(3) 29.250(3) 15.300(3)	110.67(1)	FeC_4P	OC 1.787(16) µP 2.226(1)	not given P 124.9(1)	not given	299
				FeC_4P_2	OC 1.827(13) µP 2.347(2) P 2.362(2)		not given	
$[(OC)_6Fe_2\{\mu-(PPh_2)_2C_6H_2.(PPh_2)_2\}]\cdot 2thf$	m - 2	13.639(2) 15.335(3) 14.530(2)	97.03(1)	FeC_3P_2	OC 1.770(7,11) P 2.217(2,5)	not given	C,C 98.7(3,13.2) P,P 84.4(1) C,P 96.0(2,13.4) 138.7(2) 168.6(2)	300
$[(OC)_8Fe_2\{\mu-\eta^2-PCH.(SiMe_3)_2\}_2]$ (red brown)	tr P-1 1	9.090(2) 9.551(3) 11.830(4)	109.74(2) 91.73(2) 115.52(2)	FeC_4P	OC not given P 2.226(1)	not given	not given	301

Table 5. Continued 34

Compound	Cryst. cl. / Space group / Z	a, b, c	α, β, γ	Chromophore	Fe–L	Fe–Fe	L–Fe–L	Ref.
[(OC)$_3$Fe{μ-dppt}]$_2$· 3CH$_2$Cl$_2$ (yellow)	m C2/c 4	27.170(8) 14.458(2) 8.255(3)	 102.88(2) 	FeC$_3$P$_2$	OC 1.765(9,4) P 2.211(2,5)	not given	C,C 120.0(4,2.5) P,P 177.78(9) C,P 90.0(3,2.4)	71b
[(OC)$_8$Fe$_2${μ-P$_2$(N. (SiMe$_3$)$_2$)$_2$}] (orange red) (at 140 K)	tr P-1 1	8.598(2) 9.983(2) 10.964(3)	71.38(2) 85.23(2) 68.37(2)	FeC$_4$P	OC not given P 2.232(1)	not given	not given	302
[(OC)$_6$Fe$_2${μ-ptba}] (yellow)	m P2$_1$/c 4	21.413(3) 19.256(3) 25.817(4)	 94.03(2) 	FeC$_3$P$_2$	OC 1.755(9,24) P 2.221(2,7)	7.060(2)	P,P 99.5(1,2)	303

a Where more than one chemically equivalent distance or angle is present, the mean value is tabulated. The first number in parenthesis is e.s.d., the second is maximum deviation from the mean value.

b The chemical indentity of coordinated atom/ligand is specified in these columns.

c There are two crystallographically independent molecules.

Table 6. Crystallographic and Structural Data for Trinuclear Iron Carbonyls[a].

COMPOUND (COLOUR)	CRYST. CL. SPACE GR. Z	a[Å] b[Å] c[Å]	α[°] β[°] γ[°]	CHROMO-PHORE	Fe–L [Å]	Fe–Fe [Å] Fe-L-Fe [°]	L–Fe–L [°]	REF.
A : Fe_3 triangleÅ								
$[NMe_3CH_2Ph]_2[Fe_3(\mu_3\text{-}O)(CO)_9]$ (red orange)	m $P2_1/n$ 4	11.819(1) 11.402(3) 24.271(3)	93.94(2)	FeC_3O (x3)	OC[b] 1.771 μ_3O 1.892(3,3)	2.484(1,8) O[b] not given	Fe,Fe 60.0	304
$(NPr_3Et)[(OC)_6Fe_3(\mu\text{-}CCO)$. $(\mu\text{-}SO_2)_2]$ (not given)	m $P2_1/n$ 4	10.978(3) 22.359(4) 16.810(3)	90.10(2)	FeC_4S (x2)	OC 1.75(2,1) μOC 1.99(1,1) μC 1.93(1,3) μS 2.190(4,3)	2.459(2) 2.534(3,0) C 80.0(5,1.2) S 70.4(1,1)	Fe,Fe 60.0(-,1.9) -	305
				FeC_3S_2 (x1)	OC 1.78(2,1) μC 1.99(1) μS 2.202(4,1)		not given,	
$(PPh_4)_2[(OC)_9Fe_3(\mu_3\text{-}NPh)]$. $2CH_2Cl_2$ (deep red)	m $P2_1$ 2	13.056(3) 18.114(4) 13.618(3)	98.93(2)	FeC_3N (x3)	OC not given μ_3N 1.914(11,9)	2.508(3) 2.516(3,1) N 82.1(4,1)	N,Fe 49.0(3,3) Fe,Fe 60.0(1,2)	306
$ppnOC)_9Fe_3(\mu\text{-}H)(\mu_3\text{-}NPh)]$ (red)	m $P2_1/c$ 4	15.03(3) 21.22(3) 16.12(3)	106.71(2)	FeC_3N (x1)	OC not given μ_3N 1.908(15)	2.492(7,6) 2.559(7) N 82.1(7,1.7)	N,Fe 49.2(5,1) Fe,Fe 60.0(2,1.8)	306
				FeC_3HN (x2)	OC not given μH not given μ_3N 1.916(18,2)	2.559(7) N 83.8(7)	N,Fe 48.8(6,7) Fe,Fe 59.1(2,2)	
$(OC)_9Fe_3(\mu_3\text{-}CO)(\mu_3\text{-}NPh)$	m $P2_1/c$ 4	8.634(1) 13.026(1) 17.642(2)	97.68(1)	FeC_4N	OC 1.803(4,20) μ_3OC 2.064(4,58) μ_3N 1.909(2,7)	2.510(1) 2.562(1,9) C 76.2(1,2.5) N 83.6(1,1.6)	C,C 81.6(2,6) 131.1(2,2.4) C,N 84.2(1,1.8) Fe,Fe 60.0(-,1.7)	307a

TABLE 6. Continued 2

Compound	Crystal system, space group, Z	a, b, c (Å)	α, β, γ (°)	Chromophore	Fe–L (Å)	Fe–Fe (Å), Fe–L–Fe (°)	L–Fe–L (°)	Ref.
(ppn)[Fe_3(μ-CO)$_2$($μ_3$-NH)(NO). $(CO)_6$] (colorless)	tr P-1 2	11.949(4) 13.789(4) 14.278(3)	112.35(2) 90.66(3) 93.51(3)	FeC_4N (x2)	OC 1.780(8,22) μOC not given $μ_3$HN 1.865(5,4)	2.548(1,1) $μ_3$N 85.8(2,4)	not given	307b
				FeC_2N_2 (x1)	OC not given μOC 1.792(7,10) $μ_3$HN 1.888(5) ON 1.638(6)		N,N 113.9(2)	
$H_2(OC)_9Fe_3$($μ_3$-CO)($μ_3$-NPh) (not given)	m $P2_1/a$ 4	18.343(2) 9.072(1) 11.451(1)	 102.03(1)	FeC_3N (x3)	OC 1.796(6,42) $μ_3$N 1.910(4,28)	2.510(1) 2.575(1,2) N 83.9(2,1.0)	C,C 95.2(2,5.1) C,N 102.2(2,7.7) 151.3(2,9.2) Fe,Fe 60.0(-,1.3)	307a
$(OC)_9Fe_3$(μ-H){$μ_3$-NCHMe} (not given) (at 115K)	tr P-1 2	13.620(2) 7.822(2) 7.554(1)	72.19(1) 92.7(1) 89.60(1)	FeC_4N (x1)	OC 1.795(4,7) NC 2.223(4) μN 1.960(3)	2.549(1,3) 2.5879(8) N 83.2(1,1) H 103 3 (34)	C,C 95.8(2,3.0) 162.7(2) C,N 36.1(1) 117.0(2,9) 126.6(2) Fe,Fe 60.0(-,5.6)	308
				FeC_3HN (x1)	OC 1.801(4,17) μH 1.71(6) μN 1.877(3)		C,C 95.7(2,3.6) C,H 82.9(21,5.2) 169.5(21) C,N 104.8(2,2.0) 149.1(2) 81.7(21)	
				FeC_4H (x1)	OC 1.806(4,15) μH 1.59(6)		C,C 95.9(2,1.9) C,H 82.7(24,5.4) 171.6(23)	
($AsPh_4$)[Fe_3(μ-CO)($μ_3$-HBCl). $(CO)_9$] (not given)	tr P-1 2	9.534(3) 13.305(4) 15.225(4)	104.39(2) 103.56(2) 99.23(2)	FeC_4B (x2)	OC not given μOC not given $μ_3$B 2.053(7,7)	2.571(2,31) not given	not given	309
				FeC_3B (x1)	OC not given $μ_3$B 2.137(6)			

TABLE 6. Continued 3

$(OC)_8Fe_3(\mu\text{-}CO)_2\{P(OMe)_3\}_2$[c]	m	10.749(4)		FeC_5	OC 1.817(16,17)	2.520(23)	C,C 91.3(8,7.4)	310
(dark green)	$P2_1/n$	21.776(15)	109.69(5)	(x1)	μOC 1.908(12)	2.604(27,31)	174.3(9,2.7)	
	4	11.873(10)			2.103(16)	C 78.4(5,1)		
				FeC_4P	OC 1.825(16,12)		C,C 88.4(8,4.4)	
				(x1)	μOC 1.992(24,97)		172.2(8,2)	
					P 2.166(5)		C,P 94.2(9,7.0)	
				FeC_3P	OC 1.808(15,6)		C,C 92.6(7,1.1)	
				(x1)	1.943(2)		173.6(6)	
					P 2.184(6)		C,P 94.3(5,6.2)	
				FeC_5	OC 1.817(16,17)	2.533(3)	C,C 91.9(10,11.8)	
				(x1)	μOC 1.908(10,4)	2.684(3,2)	171.0(10,1.9)	
					2.110(10)	C 78.5(6,0)		
				FeC_4P	OC 1.825(16,12)		C,C 89.6(8,7.2)	
				(x1)	μOC 1.988(22,102)		170.3(10)	
					P 2.166(5)		C,P 169.9(10,4)	
				FeC_3P	OC 1.808(16,16)		C,C 92.6(2,1.1)	
				(x1)	1.943(2)		173.6(6)	
					P 2.184(6)		C,P 94.3(5,6.2)	
$Fe_3\{\mu_3\text{-}NC(CH_2)_2Me\}(CO)_9$	m	9.003(1)		FeC_4N	OC 1.800(2,16)	2.4818(7)	C,C 96.1(1,2.7)	311
(not given)	$P2_1/m$	11.226(2)	101.33(1)	(x2)	μNC 2.076(2)	2.6257(5,0)	106.26(7)	
	2	8.986(2)			μ_3N 1.974(2)	C 73.41(8)	153.05(8)	
						N 88.17(7)	C,N 102.10(7)	
							128.1(1,1.0)	
							Fe,Fe 60.0(1,3.6)	
				FeC_3N	OC 1.799(3,1)		C,C 94.30(9)	
				(x1)	μ_3N 1.795(2)		C,N 104.99(7)	
							149.9(1)	

TABLE 6. Continued 4

Compound	Cryst. cl. Space gr. Z	a, b, c	β	Chromophore	Fe–L	Fe–Fe	Angles	Ref.
$Fe_3(\mu_3\text{-}\eta^2\text{-}NCPh)(CO)_8$. $\{P(OMe)_3\}$ (brown)	m $P2_2/c$ 4	9.336(2) 9.663(2) 27.415(5)	97.78(3)	FeC_4N (x2)	OC not given μ_3NC 2.042(4,45) μ_3N 1.966(3,10)	2.487(1) 2.636(1,10) not given	not given	312
				FeC_2NP (x1)	OC not given μ_3N 1.779(3) P 2.154(1)			
(ppn)$[Fe_3(\mu\text{-}H)(\mu_3\text{-}CO)(\mu\text{-}CO)$. $(CO)_9]$ (not given)	m $P2_1/a$ 4	15.495(3) 16.437(4) 18.127(3)	100.18(2)	FeC_5H (x2)	OC 1.767(9,16) μ_3OC 2.267(11,1) μOC 1.957(7,14) μH 1.53(4,15)	2.555(1) 2.605(1,8) not given	Fe,Fe 60.0(1,1.2)	313
				FeC_4 (x1)	OC 1.775(9,56) μ_3OC 1.765(11)		not given	
$(OC)_9Fe_3\{\mu\text{-}dab\}$ (not given)	m $P2_1/n$ 4	8.5032(6) 23.021(2) 9.6129(8)	102.212(6)	FeC_3N (x2)	OC 1.799(4,30) μN 1.900(2,3)	2.577(1,5) 2.612(1) N 83.4(1,2)	C,N 99.5(1,4.1) 158.5(1,6) Fe,Fe 60.0(1,9)	314
				FeC_3N_2 (x1)	OC 1.784(4,11) μN 1.970(2,5)		N,N 41.1(1)	
$(NBu_4)_2[(OC)_9Fe_3(\mu_3\text{-}S)]$ (brown)	tr P-1 2	10.969(2) 12.472(3) 20.363(5)	100.89 97.12 108.07	FeC_3S (x3)	OC 1.555(28,9) 1.745(19,65) μ_3S 2.195(5,10)	2.592(3,14) S 72.4(2,5)	C,C 101.8(8,10.1) C,S 102.6(5,5.1) 141.7(6,2.1) Fe,Fe 60.0(1,6)	315
$(NEt_4)_2[(OC)_9Fe_3(\mu\text{-}CO)_2]$ (dark brown)	or Pcmb 4	11.589(4) 14.979(4) 19.527(5)		FeC_5 (x2)	OC 1.752(15,32) μOC 1.963(11) 2.212(14)	2.593(3,6) 2.603(3) C 78.0(6,5.9)	C,C 76.8(4) 94.7(7,8.7) 128.2(6,2.8) 171.9(4) Fe,Fe 60.0(-,3)	316
				FeC_4 (x1)	OC 1.763(17,37) μOC 1.850(13)		C,C 99.0(5,4.6) 157.0(6)	

TABLE 6. Continued 5

$[Fe_3(\mu_3\text{-}S)(\mu\text{-}CO)(PC_2Pr^iPh).$ $(CO)_8]$ (brown)	m $P2_1/c$ 4	17.078(4) 10.031(2) 18.290(5)	110.67(2)	FeC_3S (x1)	OC not given μ_3S 2.119(1)	2.598(1,22) not given	not given	317
				FeC_3SP (x1)	OC not given μOC 1.895(5) μ_3S 2.196(1) P 2.238(2)		not given	
				FeC_3S (x1)	OC not given μOC 2.142(6) μ_3S 2.201(1)		not given	
$(PPh_4)[Fe_3(CO)_{11}]$ (burgundy) (at 213 K)	m $P2_1/n$ 4	11.313(2) 12.966(3) 23.682(5)	91.380(9)	FeC_3 (x1)	OC 1.885(12)	2.503(2) 2.657(2,23)	Fe,Fe 60.0(1,3.8)	306 318
				FeC_4 (x2)	OC 1.885(12)		C,Fe 67.4(4) 88.1(3)	
$(ppn)_2[(OC)_9Fe_3(\mu_3\text{-}Se)]$ (red)	m Cc 4	29.393(6) 14.724(3) 21.477(4)	123.65(3)	FeC_3Se (x3)	OC 1.75(2,4) μ_3Se 2.322(3,4)	2.610(3,13) Se 68.38(8,37)	C,C 99.1(8,2.6) C,Se 103.0(6,4.5) 145.8(5,2.3) Fe,Fe 60.0(-,6)	319

TABLE 6. Continued 6

$(OC)_9Fe_3(\mu\text{-}H)\{\mu_3\text{-}N(H)CMe\}$ (not given) (at 115 K)	tr P-1 2	9.276(3) 9.190(2) 9.099(2)	92.40(2) 90.79(2) 83.45(3)	FeC_4N (x1)	OC 1.792(2,5) μNC 2.096(2) μN 2.001(2)	2.542(1,2.5) 2.7537(8) C 79.07(7) N 79.58(6) 64.39(5) H 107.3(20)	Fe,Fe 60.0(-,2.6) C,C 95.7(1,3.9) 130.0(1,2.5) C,N 38.22(6) 98.23(8) 128.1(1,6.8)	308
				FeC_3HN (x1)	OC 1.806(2,8) μH 1.72(3) μN 1.931(2)		C,C 95.2(1,3.2) C,H 81.0(12,5) 171.7(12) C,N 98.6(1,3.8) 156.83(7) N,N 92.6(12)	
				FeC_4H (x1)	OC 1.801(2,27) μNC 1.933(2) μH 1.70(4)		C,C 95.5(1,6.1) 162.28(8) C,H 36.6(12) 94.4(12)	
$(ppn)[Fe_3(\mu_3\text{-}SO)(CO)_9]$ (red)	m Cc 4	29.333(5) 14.481(4) 25.048(1)	133.44(1)	FeC_3S (x3)	OC 1.774(7,22) μ_3S 2.132(2,7)	2.617(1,9) S 75.7(-,4)	Fe,Fe 60.0(-,4)	320
$(OC)_8Fe_3(\mu\text{-}CO)_2(diars)$ (dark brown)	m $P2_1/n$ 4	9.538(2) 11.071(2) 29.170(3)	99.8(1)	FeC_4 (x2)	OC 1.789(9,49) μOC 1.827(9,23)	2.599(1,3) 2.662(1) C not given	C,C 96.7(4,2.8) Fe,Fe 60.0(-,1.6)	321
				FeC_4As (x1)	OC 1.788(7,8) μOC 2.234(4,13) As 2.337(1,2)		C,C 178.9(3) As,As 86.45(4)[d] C,As 90.3(2,8)	
$(PPh_4)_2[Fe_3(\mu_3\text{-}Se)(CO)_9]$ (red brown)	tr P-1 2	11.580(4) 12.369(3) 19.707(4)	88.18(2) 81.78(2) 71.32(2)	FeC_3Se (x3)	OC 1.756(10,42) μ_3Se 2.317(2,2)	2.620(2,1) Se 68.8(1,1)	Se,Fe 55.6(1,1) Fe,Fe 60.0(1)	322
$Fe_3(\mu\text{-}H)(\mu_3\text{-}BH_4)(CO)_9$ (not given)	tr P-1 2	8.823(5) 11.523(5) 7.850(4)	98.89(4) 109.82(4) 87.92(4)	FeC_3HB (x3)	OC not given μH not given μ_3B 2.168(9,31)	2.596(2,6) 2.674(2) not given	not given	323

TABLE 6. Continued 7

$(NEt_4)_2[Fe_3(\mu_3\text{-}Se)(CO)_9]$ (dark green)	m $P2_1/c$ 4	14.210(3) 11.547(3) 19.831(2)	 90.02(2)	FeC_3Se (x3)	OC 1.68(8,8) μ_3Se 2.30(1,1)	2.623(10,7) Se not given	not given	324
$(ppn)[HFe_3(\mu_3\text{-}SO_2)(CO)_9$ (red)	m I2/a 8	36.565(9) 13.756(2) 17.648(4)	 90.87(2)	FeC_3S (x2)	OC 1.78(1,2) μ_3S 2.141(4,3)	2.625(3,24) S 76.2(1,10.0)	Fe,Fe 60.0(-,9)	320
				FeC_3OS (x1)	OC 1.76(2,7) O 2.07(1) μ_3S 2.602(5)		O,S 38.6(3)[e]	
$(PPh_4)_2[Fe_3(\mu_3\text{-}Te)(CO)_9]$ (red)	m $P2_1/n$ 4	11.034(3) 22.214(8) 21.980(8)	 97.72(3)	FeC_3Te (x3)	OC 1.747(26,28) μ_3Te 2.497(4,4)	2.625(4,9) Te 63.1(1,5)	C,C 100.9(11,3.5) C,Te 103.2(8,8.0) 140.9(6,3.5) Fe,Fe 60.0(1,5)	325
$Fe_3(\mu\text{-}CO)(\mu_3\text{-}CNCF_3)(CO)_9\cdot$ (PMe_3) (not given)	or Pbca 8	15.294(12) 15.852(9) 18,877(2)		FeC_5 (x2)	OC 1.68-1.81(3) μOC 1.97(3,10) μ_3C 1.90(3,1)	2.518(7) 2.679(7,8) OC 79(1) C 83(1)	Fe,Fe 60.0(2,3.9)	326
				FeC_3P (x1)	OC 1.68-1.81(3) Me_3P 2.23(1)			
$Fe_3(\mu\text{-}CO)(\mu\text{-}CNCF_3)(CO)_9\cdot$ $\{P(OMe)_3\}$ (not given)	or Pbca 8	15.520(4) 16.199(5) 19.626(5)		FeC_5 (x2)	OC 1.77-1.80(2) μOC 1.99(1,10) μC 1.95(1,3)	2.512(2) 2.683(2,2) OC 78.1(1) C 80.4(5)	Fe,Fe 60.0(1,4.2)	326
				FeC_3P (x1)	OC 1.77-1.80(2) P 2.172(3)			
$Fe_3(\mu\text{-}CO)(\mu\text{-}CNCF_3)(CO)_9\cdot$ $\{P(OMe)_3\}$ (not given)	tr P-1 2	8.400(3) 11.735(5) 13.374(6)	69.42(3) 90.37(3) 75.65(3)	FeC_5 (x2)	OC 1.74-1.84(3) μOC 1.95(2,5) μC 1.93(2,2)	2.529(4) 2.694(4,15) OC 80.9(9) C 82.0(9)	Fe,Fe 60.0(1,4)	326
				FeC_3P (x1)	OC 1.73-1.84(3) P 2.178(7)			

TABLE 6. Continued 8

Compound	Cryst. cl. / Space gr. / Z	a, b, c	β	Chromophore	Fe–L	Fe–Fe / Fe–X–Fe	L–Fe–L	Ref.
[$Fe_3(\mu\text{-}SC_4H_8)_2(\mu\text{-}CO)_2(CO)_6$] (red black)	m C2/c 4	15.251(3) 8.402(2) 17.445(4)	104.91(2)	FeC_4S (x3)	OC 1.79(1,1) µOC 2.16(1,40) µS 2.206(2)	2.628(2,17) S 73.8(1)	C,C 91.8(4,7.1) 101.3(5) 173.6(4) C,S 79.7(2) 101.3(3,5.0) 152.9(3)	327
$(NEt_4)_2[(OC)_9Fe_3(\mu_3\text{-}Te)]^c$ (dark red)	m $P2_1/n$ 8	24.79(2) 11.041(2) 25.209(11)	110.49(5)	FeC_3Te (x3)	OC 1.76(1,3) μ_3Te 2.486(3,10)	2.631(3,7) Te 63.89(8,23)	C,C 100.6(8,3.9) C,Te 100.2(4,2.6) 147.8(5,1.4) Fe,Fe 60.0(-,3)	319
				FeC_3Te (x3)	OC 1.75(1,10) μ_3Te 2.493(3,29)	2.630(3,3) Te 63.66(7,33)	C,C 101.5(6,6.4) C,Te 100.9(4,15.7) 142.4(4,10.5) Fe,Fe 60.0(-,2)	
$(OC)_{10}Fe_3(\mu\text{-}CO)_2$ (not given) (at 110 K)	m $P2_1/n$ 2	8.174(2) 11.190(2) 8.722(3)	96.90(2)	FeC_5 (x2)	OC 1.82 µOC 2.00(1,5)	2.540(1) 2.679(1,4) C not given	not given	328
				FeC_4 (x1)	OC 1.82			
(at 160 K)	m $P2_1/n$ 2	8.221(4) 11.128(3) 8.760(4)	96.99(3)	FeC_5 (x2)	OC 1.82 µOC 2.02(3,6)	2.545(2) 2.679(4,5) C not given	not given	328 329
				FeC_4 (x1)	OC 1.82			
(at 250 K)	m $P2_1/n$ 2	8.304(2) 11.231(4) 8.822(3)	96.89(4)	FeC_5 (x2)	OC 1.82 µOC 2.04(2,7)	2.551(2) 2.676(1,3) C not given	not given	328
				FeC_4 (x1)	OC 1.82			

TABLE 6. Continued 9

(at 320 K)	m $P2_1/n$ 2	8.375(2) 11.330(2) 8.882(4)	 97.00(3)	FeC_5 (x2)	OC 1.82 μOC 2.12(3,17)	2.554(2) 2.679(1,3) C not given	not given	328
				FeC_4 (x1)	OC 1.82			
$(OC)_{10}Fe_3(\mu\text{-}CO)_2$ (charcoal black)	m $P2_1/n$ 2	8.36(1) 11.30(2) 8.87(1)	 97.5(10)	FeC_5 (x2)	OC 1.85(3,7) μOC 2.21(2,13)	2.560(6) 2.637(7,5) C 70.7(7,2.0)	C,C 83.4(7,3.6) 101.6(9,11.2)] 168.6(8,1.5) Fe,Fe 60.0(2,2.8)	330
				FeC_4 (x1)	OC 1.86(2,17)		C,C 92.3(7,1.6) 115.8(7) 172.1(7)	
$(OC)_{10}Fe_3(\mu\text{-}CO)(\mu\text{-}CNCF_3)$[c] (dark green)	tr P-1 4	14.607(1) 14.289(2) 8.967(1)	92.40(2) 92.70(2) 90.66(1)	FeC_5 (x2)	OC 1.79(1,3) μOC 1.97(1,3) μC 1.96(1,2)	2.526(2) 2.689(3,4) OC 79.5(4,2) C 80.0(4)	Fe,Fe 60.0(-,3.7)	331
				FeC_4 (x1)	OC 1.79(1,3)		not given	
				FeC_5 (x2)	OC 1.79(1,3) μOC 2.00(1,2) μC 1.96(1,3)	2.536(2) 2.688(3,4) OC 78.7(5) C 80.7(4)	Fe,Fe 60.0(-,4.0)	
				FeC_4 (x1)	OC 1.79(1,3)		not given	
$Fe_3(\mu\text{-}CO)_2(PMe_2Ph)_3(CO)_7$ (not given)	or Pbca 8	16.74(4) 18.75(4) 23.43(5)		FeC_3P (x2)	OC 1.78(3,6) μOC 2.00(3,4) P 2.234(9,2)	2.540(7) 2.688(7,1) C not given	Fe,Fe 60.0(2,3.6)	332
				FeC_3P (x1)	OC 1.77(2,7) P 2.242(9)			

TABLE 6. Continued 10

$(OC)_{10}Fe_3(\mu\text{-}CO)_2$ (not given)	m $P2_1/n$ 2	8.359(2) 11.309(2) 8.862(2)	 97.00(2)	FeC_5 (x2)	OC 1.80(3,3) μOC 2.05(4,16)	2.558(1) 2.680(2,3) C not given	C,C 91.0(12,6.0) 175.6(13,2.1) Fe,Fe 60.0(-,3.0)	333
				FeC_4 (x1)	OC 1.85(4,4)		C,C 93.9(12,6.0) 172.0(12)	
$Fe_3(\mu\text{-}CO)(\mu_3\text{-}\eta^2\text{-}OPPr^i)(CO)_9$ (not given)	tr P-1 2	9.105(7) 9.462(8) 11.406(8)	97.22(6) 97.81(6) 101.01(6)	FeC_4P (x2)	OC 1.803 μOC 1.99(1,0) μ_3P 2.187(4,8)	2.629(3) 2.647(3,1) C 82.7(3) 70.1(1,3.8)	Fe,Fe 60.0(1,5)	334
				FeC_3OP (x1)	OC 1.803 O 2.051(7) μ_3P 2.515(4)			
$Fe_3(\mu\text{-}CO)(\mu_3\text{-}\eta^2\text{-}OPBu^t)(CO)_9$ (not given)	tr P-1 2	9.239(4) 9.625(5) 11.720(5)	101.17(4) 9.625(5) 11.720(5)	FeC_4P (x2)	OC 1.811 μOC 2.013(7,18) μ_3P 2.211(2,2)	2.637(2) 2.668(2,9) C 81.8(3) P 69.7(1,3.5)	Fe,Fe 60.0(1,8)	334
				FeC_3OP (x1)	OC 1.811 O 2.026(5) μ_3P 2.540(2)			

TABLE 6. Continued 11

$Fe_3(\mu\text{-}H)\{\mu_3\text{--}\eta^2\text{-}O(Et)PBu^t\}$. $(CO)_9$[f] (not given)	tr P-1 6	9.583(5) 17.858(8) 18.656(8)	96.13(3) 100.52(3) 91.32(3)	FeC_3HP (x2)	OC 1.777 μH 1.68(5,5) μ_3P 2.154(2,3	2.641(2,4) 2.688(2) H 107(3) P 70.6(1,6.6)	Fe,Fe 60.0(1,1.2)	334
				FeC_3OP (x1)	OC 1.777 O 2.084(4) μ_3P 2.751(2)		not given	
				FeC_3HP (x2)	OC 1.780 μH 1.66(5,4) μ_3P 2.158(2,4)	2.643(2,9) 2.692(2) H 108(3) 70.5(1,6.7)		
				FeC_3OP (x1	OC 1.780 O 2.091(3) μ_3P 2.575(2)		not given	
				FeC_3HP (x2)	OC 1.772 μH 1.68(5,4) μ_3P 2.155(2,3)	2.645(2,13) 2.697(2) H 106(3) P 70.7(1,6.8)		
				FeC_3OP (x1)	OC 1.772 O 2.077(3) μ_3P 2.537(2)		not given	
$[(OC)_7Fe_3(\mu\text{-}CO)(\mu_3\text{-}Se)(PPh_3)_2]$. $0.5CH_2Cl_2$ (not given)	tr P-1 2	13.405(6) 14.303(6) 12.344(5)	79.72(2) 77.01(2) 71.04(2)	FeC_3PSe (x2)	OC 1.783(9,8) μOC 1.987(6,13) Ph_3P 2.259(2,9) μ_3Se 2.332(1,7)	2.642(1,26) C not given Se 69.3(1,1.0)	C,C 84.3(3,2.1) 106.4(3,9) C,P 89.3(2,3.3) 168.3(2,8) C,Se 118.8(3,16.6) P,Se 88.0(1,6)	335
				FeC_3Se (x1)	OC 1.792(10,18) μ_3Se 2.316(2)		C,C 93.5(4,4.0) C,Se 105.0(3,1.3) 151.6(2)	

TABLE 6. Continued 12

Compound	Crystal class, space group, Z	a, b, c	β	Chromophore	Fe–L	Fe–Fe	L–Fe–L	Ref.
$(NPr^i_2)[Fe_3(\mu_3\text{-}SBu^t)(CO)_9]$ (red)	m $P2_1/c$ 4	8.574(1) 15.923(3) 18.841(3)	 95.33(5)	FeC_3S (x3)	OC 1.745 μ_3S 2.120(2,2)	2.644(1,10) S not given	C,C 97.3-101.1	336
$(OC)_8Fe_3(\mu\text{-}CO)_2\{P(OMe)_3\}$. (Bu^tCN)[c] (black)	m $P2_1/n$ 8	11.514(3) 28.991(7) 16.498(5)	 100.13(3)	FeC_5 (x1)	OC 1.798(10,23) μOC 2.021(9,46)	2.548(2) 2.694(2,2) C 79.4(3,4)	C,C 91.9(4,6.6) 171.5(4,2.2) Fe,Fe 60.00(4,3.56)	337
				FeC_4P (x1)	OC 1.789(10,6) μOC 1.967(9,33) P 2.171(3)		C,C 89.9(4,6.6) 172.8(4,3.9) C,P 93.2(3,5.9)	
				FeC_4 (x1)	OC 1.799(10,7) NC 1.892(9)		C,C 93.8(4,8.1) 176.2(4)	
				FeC_5 (x1)	OC 1.800(11,34) μOC 2.020(10,10)	2.548(2) 2.694(2,2) C 79.4(4,1)	C,C 91.8(4,6.6) 171.8(4,9) Fe,Fe 60.00(4,3.56)	
				FeC_4P (x1)	OC 1.784(11,10) μOC 1.968(9,14) P 2.174(3)		C,C 89.6(4,5.6) 173.3(4,0) C,P 94.7(3,5.0)	
				FeC_4 (x1)	OC 1.792(10,8) NC 1.882(9)		C,C 89.2(4) 175.9(4)	

TABLE 6. Continued 13

$(OC)_9Fe_3(\mu\text{-}CO)_2(PPh_3)$ [c] (dark green)	m C2/c 2	37.14(3) 12.26(1) 26.05(2)	93.96(15)	FeC_4 (x1)	OC 1.74(5,4)	2.568(8) 2.684(9,19) C 83(2,1)	C,C 93(2,2) 172(2)	338
				FeC_4P (x1)	μOC 1.93(4,7) OC 1.71(4,4) P 2.24(1)		C,C 94(2,3) 169(2,2) C,P 94(2,3)	
				FeC_5 (x1)	μOC 1.97(4,7) OC 1.73(5,2)		C,C 93(2,7) 170(2,1)	
				FeC_5 (x2)	μOC 1.96(5,4) OC 1.73(6,8)	2.558(9) 2.707(9,4) C 82(2,1)	C,C 94(2,6) 172(2,1)	
				FeC_3P (x1)	OC 1.71(5,3) P 2.25(1)		C,C 94(2,2) 169(2) C,P 94(2,2)	
$Fe_3(\mu\text{-}CO)(\mu_3\text{-}PBu^t)CO_9$ (orange red)	or Pnma 4	14.87(2) 13.54(2) 9.671(6)		FeC_{4P} (x2)	OC 1.77 μOC 1.97(1) μ_3P 2.174(4)	2.646(4,23) C not given P not given	not given	339
				FeC_3P (x1)	OC 1.77 μ_3P 2.124(5)			
$[Fe_3(\mu_3\text{-}S_6H_{11}S)(CO)_9((\mu\text{-}H)]$ (red)	tr P-1 2	8.174(8) 8.182(7) 17.55(2)	93.07(8) 82.99(8) 118.27(6)	FeC_3S (x1)	OC 1.762 μ_3S 2.123(3)	2.653(3,16) S not given	C,C 95.1-102.8	336
				FeC_3HS (x2)	OC 1.762 μH 1.73(9,10) $_3$S 2.134(3,4)		not given	

TABLE 6. Continued 14

Compound								
$[Fe_3(\mu_3\text{-}SBu^t)(PPh_3)(CO)_8(\mu\text{-}H)]$ (brown black)	m $P2_1/c$ 4	9.095(8) 15.51(2) 22.61(2)	97.68(7)	FeC_3S (x1)	OC 1.778 μ_3S 2.140(2)	2.618(1,8) 2.732(1) S not given	C,C 92.5-98.2	336
				FeC_3HS (x1)	OC 1.778 μ_3S 2.157(2) μH 1.76(6)		not given	
				FeC_2HPS (x1)	OC 1.778 μH 1.68(7) P 2.237(2) μ_3S 2.148(2)		not given	
$[N(PPh_3)_2][Fe_3(\mu\text{-}H)(\mu\text{-}CO).$ $(CO)_{10}]$ (dark brown)	m $P2_1/c$ 4	18.158(13) 16.512(10) 15.492(13)	99.62(7)	FeC_4H (x2)	OC 1.768(13,37) μOC 1.930(11,8) μH 1.79(7,1)	2.572(2) 2.686(2,13) C 83.5(4) H 92(3)	C,C 93.6(5,5.2) 165.6(5,6) Fe,Fe 60.0(1,2.8)	340
				FeC_4 (x1)	OC 1.768(14,32)		C,C 96.2(6,7.5) 167.6(5)	
$Fe_3(\mu_3\text{-}Bu^t)(PPh_3)_2(CO)_7(\mu\text{-}H)$ (brown black)	tr P-1 2	11.96(1) 14.04(1) 17.20(2)	106.09(6) 110.20(6) 88.13(6)	FeC_3S (x1)	OC 1.743 μ_3S 2.120(6)	2.468(4,23) 2.782(4) S not givenn	C,C 88.8-99.7	336
				FeC_2HPS (x2)	OC 1.743 μH 1.63(-,15) P 2.267(6,21) μ_3S 2.135(5,5)			
$Fe_3(\mu\text{-}CO)_2(CO)_8(CHBu^t)_2$ (not given)	m $P2_1/n$ 4	11.111(4) 23.023(2) 11.336(4)	114.23(3)	FeC_5 (x2)	OC not given μOC 1.986(8,11)	2.553(2) 2.699(1,3) not given	Fe,Fe 61.8(1,1)	341
				FeC_4 (x1)	OC not given NC 1.836(10,66)		C,C 94.1(4,7.9) Fe,Fe 56.5(1)	

TABLE 6. Continued 15

(NEt$_3$H)[(OC)$_{10}$Fe$_3$(μ-H)(μ-CO)] (dark)	tr P-1 2	10.04 10.46 11.57	94.11 90.43 104.62	FeC$_4$H (x2)	OC not given μOC not given μH not given	2.557(3) 2.690(3,6)	not given	342
				FeC$_4$ (x1)	OC notgiven			
(Cx$_2$NH$_2$)[HFe$_3$(μ-CO)(CO)$_{10}$]. Et$_2$O (not given)	m P2$_1$/a 4	17.449(5) 16.345(5) 13.466(4)	118.45(2)	FeC$_4$ (x2)	OC 1.78(1,6) μOC 1.901(7,9)	2.594(2) 2.682(3,2) C 86.0(4)	C,C 94.7(6,6.2) 165.7(4,1.7) Fe,Fe 60.0(1,2.2)	343
				FeC$_4$ (x1)	OC 1.790(2,4)		C,C 95.8(7,8.2) 168.0(6)	
(ppn)[Fe$_3$(μ-H)(μ-SO$_2$)(CO)$_9$][c] (red)	tr P-1 4	8.622(3) 14.046(3) 15.289(3)	69.11(2) 77.18(2) 74.74(2)	FeC$_3$HS (x2)	OC not given μH 1.65(8,6) μS 2.091(2,5)	2.629(2) 2.666(2,9) not given	not given	344
				FeC$_3$O (x1)	OC not given O 2.078(4)			
				FeC$_3$HS (x2)	OC not given μH 1.64(8,1) μS 2.089(2,12)	2.639(2,1) 2.680(2) not given	not given	
				FeC$_3$O (x1)	OC not given O 2.070(4)			
[(OC)$_7$Fe$_3$(μ-CO)$_2${P(OMe)$_3$}$_3$] (black)	m P2$_1$/c 4	9.382(4) 17.974(4) 18.950(4)	99.57(2)	FeC$_4$P (x2)	OC 1.786(12,28) μOC 2.002(9,79) P 2.162(3,3)	2.556(2) 2.705(2,0) C not given	C,C 89.8(4,7.1) 172.7(4,1.4) C,P 93.3(4,3)	345
				FeC$_3$P (x1)	OC 1.796(10,7) P 2.172(2)		C,C 93.7(4,2.5) 172.1(4) C,P 94.0(3,6.1)	

TABLE 6. Continued 16

$Fe_3)\mu$-CO)(μ_3-PPh)$(CO)_8$. {$P(OMe)_3$} (dark red)	M $P2_1/c$ 4	9.593(4) 24.047(1) 14.871(6)	132.94(2)	FeC_4P (x1)	OC 1.77 μOC 1.93(2) μ_3P 2.180(6)	2.656(4,42) P not given	not given	339
				FeC_3P_2 (x1)	OC 1.77 μOC 1.93(2) μ_3P 2.147(3) P 2.165(7)			
				FeC_3P (x1)	OC 1.77 μ_3P 2.118(3)			
$[Fe_3(\mu$-$SPr^i)(\mu$-$H)(CO)_9]$ (not given)	m $P2_1/c$ 4	9.59(1) 9.88(1) 19.59(2)	96.11(7)	FeC_3HS (x2)	OC 1.79(1,3) μH 1.81(13,2) μS 2.147(3,8)	2.657(2,21) S 76.61(9,57) H 96(6)	C,C 97.3(5,5.8) C,H 67(4,10) 89.5(4,5) 163(4,10) C,S 104.3(4,5.7) 145.1(3,2) H,S 93.5(4,5)	346
				FeC_3S (x1)	OC 1.80(1,2) μS 2.136(2)		C,C 98.1(5,1.8) C,S 105.7(3,3) 142.4(4)	
$[H_2(Pr^i)_2N][Fe_3(\mu$-H). (μ-CO)$(CO)_{10}$] (not given)	m $P2_1/c$ 4	9.635(4) 21.182(7) 11.897(5)	104.01(3)	FeC_4H (x2)	OC 1.784(16,61) μOC 1.916(11,41) μH 1.631(-,21)	2.579(2) 2.706(3,3) C 84.6(4) H 104.4	C,C 94.9(7,7.0) 164.3(6,2.4) C,H 60.1(-,7)	347
				FeC_4 (x1)	OC 1.768(13,20) 1.816(14,25)		C,C 95.0(7,5.4) 167.9(7)	
$Cs[Fe_3(CO)_9TeH]$ (red brown)	or Pbca 8	9.672(2) 15.816(4) 21.430(5)		FeC_3Te (x3)	OC not given μ_3Te 2.500(3,10)	2.637(3,10) 2.736(4) Te 58.42(9)	not given	348

TABLE 6. Continued 17

Compound	Crystal class, space group, Z	a, b, c	α, β, γ	Chromophore	M–L	M–M, M–L–M	L–M–L	Ref.
(NBu$_4$)[Fe$_3$(μ-H)(μ$_3$-pan)(CO)$_9$] (red)	not given			FeC$_3$HP (x2)	OC 1.77(-,3) μH 1.591(2,5) μ$_3$P 2.163(3,0)	2.678(3,1) 2.718(3) H 117.3(1) P not given	Fe,Fe 60.0(1,1.0)	349
				FeC$_3$P (x1)	OC 1.77(-,3) μ$_3$P 2.145(3)			
Fe$_3$(μ-H){μ-P(Et)PhOMe-4}. (CO)$_9${P(OPh)$_3$} (red) (at 213 K)	tr P-1 2	10.605(6) 12.186(5) 15.504(9)	82.61(4) 103.45(4) 106.57(4)	FeC$_3$HP (x2)	OC 1.782(-,17) μH 1.58(6,4) μP 2.198(2,2)	2.687(2,1) 2.702(2) H 115(4) P 75.4(1)	Fe,Fe 60.0(-,4)	349
				FeC$_3$P (x1)	OC 1.782(-,17) P 2.146(2)			
Fe$_3$(μ-H)$_2${μ$_3$-PSi(Pr$^i_3$)}(CO)$_9$ (dark red) (at 123 K)	m C2/c 8	33.822(14) 14.299(14) 10.365(3)	93.33(5)	FeC$_3$HP (x2)	OC not given μH not given μ$_3$P 2.190(2,5)	2.665(2) 2.714(3,4) not given	not given	350
				FeC$_3$H$_2$P (x1)	OC not given μH not given μ$_3$P 2.241(2)			
(OC)$_9$Fe$_3${P(OPr^i)$_3$}$_3$ (black)	m P2$_1$/n 4	9.827(6) 40.936(13) 13.431(4)	107.14(4)	FeC$_3$P (x3)	OC 1.796(12,29) P 2.188(3,7)	2.699(2,14) Fe 60.0(1,4)	C,C 95.4(5,1.8) 168.3(4,1.0) C,P 93.2(4,5.0)	345

TABLE 6. Continued 18

$(OC)_8Fe_3(\mu\text{-}H)_2(\mu\text{-}PMe_2)_2$ (black)	tr P-1 2	9.487(3) 13.925(2) 8.441(2)	101.77(1) 112.83(2) 71.89(2)	FeC_3HP (x2)	OC 1.782(7,18) μH 1.57(1) μP 2.241(2,1)	2.646(2,4) 2.826(1) P 73.59(5,10) H 116(3,3)	C,C 96.2(4,1.9) C,H 91(2,7) 167(2,2) C,P 93.3(2,1.5) 165.4(2,4) H,P 79(2,2)	351
				$FeC_2H_2P_2$ (x1)	OC 1.746(6,1) μH 1.57(5,5) μP 2.177(2,2)		C,C 91.5(3) C,P 102.4(3,3.1) C,H 84(2,2) 173(2,1) H,P 79(2,2) 101(3) H,H 101(3) P,P 144.44(7)	
$(OC)_7Fe_3(\mu_3\text{-}Se)_2(PPh_3)_2$ (not given)	m $P2_1/c$ 4	14.851(6) 19.321(8) 15.016(6)	104.68(2)	FeC_2PSe_2 (x2)	OC 1.772(6,17) Ph_3P 2.255(2,3) μ_3Se 2.364(1,10)	2.715(1,16) Se 70.0(1,7) 96.4(1,5)	C,C 97.0(3,1) C,P 95.1(2,3.9) C,Se 95.5(2,7.5) 161.1(2,1.8) Se,Se 81.9(1,1) P,Se 90.2(1,1.3) 167.3(1,4)	335
				FeC_3Se_2 (x1)	OC 1.793(7,1) μ_3Se 2.366(1,0)		C,C 95.9(3,4.5) Se,Se 81.9(1) C,Se 91.3(2,2.7) 129.7(2,4.5) 169.8(2)	
$Fe_3(\mu\text{-}H)_2(\mu_3\text{-}PPh)(CO)_9$ (red) (at 193 K)	m $P2_1/c$ 4	10.033(6) 8.542(4) 23.62(3)	97.64(7)	FeC_3H_2P (x1)	OC 1.813(8,9) μH 1.49(6,9) μ_3P 2.199(2)	2.692(1) 2.728(1,1) H 117(4,6) P not given	not given	352
				FeC_3HP (x2)	OC 1.789(7,26) μH 1.72(6,1) μ_3P 2.150(2,4)			

TABLE 6. Continued 19

$(OC)_8Fe_3(\mu\text{-}H)_2(\mu\text{-}PPh_2)_2$ (dark green)	tr P-1 2	10.918(2) 11.898(2) 14.705(3)	75.02(1) 84.72(1) 70.84(1)	$FeH_2C_2P_2$ (x1)	μH 1.61(4,8) OC 1.761(4,1) μP 2.204(1,3)	2.663(1,4) 2.8264(8) H 114(1,4) P 73.2(-,1)	H,H 103(2) C,C 90.9(1) P,P 143.63(3)	353
				FeC_3HP (x2)	OC 1.788(5,16) μH 1.57(4,2) μP 2.259(1,3)		C,C 94.5(2,3.5) C,H 90(1,5) 170(1,2) C,P 95.9(1,4.1) 165.0(1,1.2) H,P 79(1,4)	
$[Fe_3(NCC_6H_4\text{-}2\text{-}Me)(\mu\text{-}CO)_2\cdot$ $(CO)_9]$ (black green)	m C2/c 4	7.748(1) 24.855(2) 12.300(2)	107.62(1)	FeC_5 (x2)	OC 1.750(23,11) μOC 1.850(24) 1.900(24,1) 2.185(27)	2.551(4) 2.809(4,22) C 81.3(8,4.4)	C,C 97.1(1.0,2.4) μC,C 89.22(0.9,10.1 μC,μC 6) 162.71(1.0,21) 77.12(1.1)	354
				FeC_3N (x1)	OC 1.782(33,53) N 1.970(20)		C,C 93.6(1.3,5.6) C,N 91.8(1.1,1.1) 176.0(1.1)	
$Fe_3(\mu\text{-}H)\{\mu\text{-}PMeBu^t\}\cdot$ $\{\mu\text{-}BuPCH_2PBu^t\}(CO)_7$ (red)	m $P2_1/c$ 4	16.654(9) 16.113(5) 10.808(3)	92.24(3)	FeC_2P_2H (x2)	OC not given μP 2.176(-,6) 2.203(-,1) μH 1.69(3,2)	2.5570(5) 2.818(-,12) P 76.1(2,5.2) H 98.7(13)	Fe,Fe 60.0(1,6.0) P,P 142.2(1,4) P,H 80.9(9,2.6)	355
				FeC_3P_2 (x1)	OC not given μP 2.265(-,5)		P,P 71.62(3)	

TABLE 6. Continued 20

Compound	Crystal class / Space group / Z	a, b, c	α, β, γ	Chromophore	Fe-L	Fe-Fe / Fe-L-Fe	L-Fe-L	Ref.
$Fe_3(\mu$-H){μ-P(Pr^i)CH_2PPr$^i_2$}. $(CO)_9$ (not given)	m $P2_1/c$ 4	8.858(3) 16.391(6) 18.526(5)	 104.22(2)	FeC_3HP (x1)	OC 1,780(5,30) μH 1.70(5) μP 2.194(1)	2.715(-,15) 2.7993(7) H 117(3) P 76.26(3)	Fe,Fe 60.0(-,2.1) C,C 95.6(2,5.4) C,H 89(1,2) 174(1) C,P 93.5(2,1.9) 166.7(1) H,P 86(2)	356
				FeC_3HP (x1)	OC 1.780(5,13) μH 1.59(5) P 2.293(1)		C,C 93.7(2,7.2) C,H 101(2,15) C,P 90.9(1,1.8) 177.4(2) H,P 80.8(1)	
				FeC_3P (x1)	OC 1.782(5,10) μP 2.1779(8)		C,C 95.8(2,5.4) C,P 96.4(1,2.9) 158.5(2)	
$(OC)_9Fe_3(\mu$-H$)_3(\mu_3$-Bi) (orange brown)	tr P-1 2	7.997(2) 9.018(2) 12.464(4)	78.83(2) 71.16(2) 68.78(2)	FeC_3H_2Bi (x3)	OC not given μH 1.76(3,0) μ_3Bi 2.639(1,2)	2.774(2,2) H 59.9(1) Bi 63.3(1)	C,C not given H,H 80(4) H,Bi 96(2)	357
B : Fe_3 isosceles triangle with two Fe-Fe bonds								
$Fe_3(\mu_3$-NPh$)_2(CO)_8${P(OMe)$_3$} (dark purple)	tr P-1 4	9.065(2) 17.474(8) 17.989(6)	87.11(4) 84.20(3) 88.35(4)	FeC_3N_2 (x2)	OC 1.789(4,13) μ_3N 1.957(2,18)	2.450(1,28) 3.079(1) N 75 105	not given	358
				FeC_2N_2P (x1)	OC 1.787(4,3) μ_3N 1.937(2,2) P 2.151(1)		not given	
[$Fe_3(\mu_3$-NMe$)_2(CO)_9$] (deep purple)	tr P-1 1	8.31(2) 8.90(2) 12.79(3)	80.48(6) 108.45(6) 111.48(6)	FeC_3N_2 (x3)	OC 1.74(3,4) μ_3N 1.93(2,4)	2.462(7,26) 3.044(8) N 79.2(7,2.0) 104.7(8,0)	C,C 94.5(13,3.3) C,N 100.1(11,12.1) 156.9(10,6.0) N,N 71.3(7,2)	360

TABLE 6. Continued 21

Compound	Crystal class Space group Z	a b c	α β γ	Chromophore	Fe–L	Fe–Fe / Fe–L–Fe	L–Fe–L	Ref.
$(OC)_9Fe_3(\mu_3\text{-}S)\{\mu_3\text{-}4\text{-}MeC_6H_4N\}$[c] (not given)				FeC_3NS (x3)	OC 1.78(3,9) μ_3N 1.92(2,2) μ_3S 2.24(2,1)	2.51(1,1) N 81.8(7,7) 112.5(7) S 68.3(2,4) 91.1(3)	C,N 100(1,6) 158(1,1) C,S 115(1,30) 168(1,3) N,S 75.5(6,5)	158
				FeC_3NS (x3)	OC 1.77(3,8) μ_3N 1.94(2,1) μ_3S 2.24(1,1)	2.52(1,1) N 80.7(7,4) 109.5(8) S 68.6(2,2) 90.7(3)	C,N 100(1,4) 162(1,4) C,S 101(1,16) 166(1,2) N,S 76.9(5,1)	
$(OC)_7Fe_3\{\mu\text{-}S(CH_2)_2S\}_2$ (dark red) (at 223 K)	tr P-1 2	6.465(6) 8.32(1) 18.36(4)	102.6(1) 83.2(1) 111.36(8)	FeS_4C (x1)	μS 2.227(4,13) OC 1.78(1)	2.551(4,14) S 59.6	S,S 81.6(1,1) 94.8(1,2.6) 168.0(9)	361
				FeC_3S_2 (x2)	OC 1.79(2,6) μS 2.246(4,14)			
$Fe_3(\mu_3\text{-}NPh)(\mu_3\text{-}PBu^t)(CO)_9$ (red brown)	or $Pbc2_1$ 4	12.109(3) 14.793(5) 12.982(3)		FeC_3NP (x3)	OC not given μ_3N 1.959(12,10) μ_3P 2.208(5,1) 2.251(4)	2.564(3,5) N 81.9(4,4) 114.3(5) P 70.2(1,2) 96.5(2)	Fe,Fe 80.0(1) N,P 72.2(3,6)	362
$[Fe_3(\mu_3\text{-}S)_2\{P(SC_6H_5)Cl_2\}(CO)_8]$ (not given)	tr P-1 2	7.571(1) 9.097(1) 17.706(2)	95.03(1) 101.79(1) 105.73(1)	FeC_3S_2 (x1)	OC not given μ_3S 2.227 2.236	2.594(-,15) 3.3601 μ_3S 98.02(4,22)	S,S 80.28(4)	363
				FeC_2S_2P (x1)	OC not given μ_3S 2.238-2.250 P 2.147(2)			
				FeC_3S_2P (x1)	OC not given μ_3S 2.23-2.25 P 2.147(2)		S,S 80.54(4)	

TABLE 6. Continued 22

Compound	System	Cell	Angles	Chromophore	Fe–L	Fe–Fe, Fe–L–Fe	L–Fe–L	Ref.
$Fe_3(\mu_3\text{-}S)_2(NHMe_2)(CO)_8$ (dark red)	m $P2_1/c$ 4	7.004(1) 13.652(2) 17.764(2)	 91.87(1)	FeC_3S_2 (x2)	OC 1.79(1,1) μ_3S 2.250(2,13)	2.607(2,38) S 71.1(1,1.4) 98.12(9)	C,C 96.8(4,2.6) S,S 79.3(1,3) C,S 93.7(3,6.6) 118.6 141.3(3) 161.7(3,1.5)	364
				FeC_2S_2N (x1)	OC 1.77(1,2) μ_3S 2.218(2,4) N 2.088(6)		C,C 93.6(4) S,S 80.72(8) C,N 95.7(3,1.7) S,N 89.7(2) 164.3(2)	
$(OC)_9Fe_3(\mu_3\text{-}S)(\mu\text{-}\eta^2\text{-}NSBu^t)$ (not given)				FeC_3S_2N (x1)	OC 1.76(2,1) 1.87(2) μ_3S 2.314(6) μS 2.227(5) μN 2.08(1)	2.614(4) μ_3S 71.2(2) 92.7(2,2.1) S 97.1(5) N 101.8(6)	C,C 94(1,6) S,S 82.4(2) C,N 153(1) C,S 163(1,13) S,N 81.6(4)	158
				FeC_3SN (x1)	OC 1.75(2,1) 1.86(2) μ_3S 2.237(6) μN 2.09(1)		C,C 95(1,9) C,S 148(1) C,N 171(1) S,N 83.1(4)	
				FeC_3S_2 (x1)	OC 1.75(2,1) μ_3S 2.253(6) μS 2.259(5)		C,C 94(1,8) S,S 83.1(2) C,S 166(1,9)	
$(OC)_9Fe_3(\mu_3\text{-}S)(\mu_3\text{-}Se)$ (dark red)	tr P-1 2	6.8346(10) 9.186(2) 13.121(3)	93.79(3) 94.21(3) 110.89(3)	FeC_3SSe (x3)	OC 1.801(3,21) μ_3S 2.290(1,7) μ_3Se 2.351(1,14)	2.620(1,7) 3.438(1) S 69.7(1,2) 97.5(1) Se 67.6(1,2) 94.3(1)	C,C 95.8(1,4.4) C,S 91.9(1,6.2) 129.9(1,3) 165.3(1,3) C,Se 94.5(1,4.1) 164.7(1,3.4) S,Se 82.8(1,5)	365

TABLE 6. Continued 23

Compound	Cryst. cl. / Space group / Z	a, b, c	α, β, γ	Chromophore	M–L	M–M / M...M	Angles	Ref.
$Fe_3(\mu_3-S)(\mu_3-SO)(CO)_9$ (red brown)	tr P-1 2	9.514(1) 10.320(1) 9.065(4)	91.38(2) 116.97(2) 97.06(1)	FeC_3S_2 (x3)	OC not given μ_3S 2.254(1,24) μ_3OS 2.167(1,55)	2.635(1,12) not given	not given	366
$[Fe_3(\mu_3-S)(\mu_3-ptp)(CO)_9]$ (dark red)	m $P2_1/n$ 4	8.558(2) 9.022(2) 27.506(6)	 95.40(2)	FeC_3PS (x3)	OC not given μ_3P 2.200(1,44) μ_3S 2.263(1,9)	2.643(1,7) 3.495(1) not given	P,S 75.1(1,6)	367
$Fe_3(\mu_3-S)(\mu_3-PPr^i)(CO)_8(CN_2H_5)$[c] (not given)	m $P2_1/n$ 8	10.229(3) 13.116(3) 29.981(7)	 90.04(2)	FeC_3SP (x2)	OC not given μ_3S 2.256(3,4) μ_3P 2.223(9,24)	2.646(2,12) S 102.3(1) P 106.0(1)	Fe,Fe 83.0(1) S.P 74.3(1,6)	368
				FeC_3SP (x1)	OC not given NC 2.016(9) μ_3S 2.250(3) μ_3P 2.189(3)		C,S 89.5(3) C,P 160.0(3) S,P 75.2(1)	
				FeC_3SP (x2)	OC not given μ_3S 2.255(3,2) μ_3P 2.229(3,23)	2.657(2,22) S 102.3(1) P 106.2(1)	Fe,Fe 82.8(1) S,P 74.2(4,4)	
				FeC_3SP (x1)	OC not given NC 2.037(10) μ_3S 2.255(3) μ_3P 2.187(3)		C,S 90.7(3) C,P 161.6(3) S,P 75.1(1)	
$Fe_3(\mu_3-Se)_2(CO)_9$ (not given) (at 153 K)	tr P-1 2	6.748(1) 9.163(1) 13.044(1)	93.97(1) 94.54(1) 110.75(1)	FeC_3Se_2 (x3)	OC 1.800(7,22) μ_3Se 2.359(1,26)	2.651(1,6) 3.520(1) not given	not given	369
$Fe_3(\mu_3-S)_2(CO)_9$ (not given)	tr P-1 2	6.682(2) 9.047(2) 12.998(3)	93.96(2) 94.42(2) 111.03(2)	FeC_3S_2 (x3)	not given	not given	not given	369
$(OC)_9Fe_3(\mu_3-S)(\mu_3-PPh)$ (red) (at 211 K)	tr P-1 6	11.49(2) 12.17(2) 22.49(3)	104.0(1) 98.6(1) 90.5(1)	FeC_3SP (x3)	OC 1.757 μ_3S 2.258(6,12) μ_3P 2.194(6,5) 2.259(5)	2.655(5,9) 3.511(6) S not given P not given	C,C 95.1(8,3.7) S,P 74.7(2,6)	370

TABLE 6. Continued 24

[$Fe_3(\mu_3\text{-}S)\{\mu_3\text{-}P(4\text{-}MeOC_6H_4)\}$. $(CO)_9$] (dark red)	m C2/c 8	21.977(8) 12.198(4) 16.609(8)	 103.20(4)	FeC_3PS (x3)	OC 1.793(6,34) μ_3P 2.205(1,52) μ_3S 2.263(1,29)	2.660(1,26) 3.499(1) μ_3S 72.25(4,85) 100.32(5) μ_3P 73.60(4,1.22) 106.54(5)	P,S 75,24(4,37)	371
$(OC)_9Fe_3(\mu_3\text{-}S)(\mu_3\text{-}Te)$ (dark red)	tr P-1 2	6.9956(8) 9.3675(14) 13.0584(16)	93.227(11) 94.031(10) 111.138(10)	FeC_3STe (x3)	OC 1.797(8,23) μ_3S 2.283(2,6) μ_3Te 2.533(1,21)	2.663(1,9) 3.546(1) S 71.3(1,3) 102.1(1) Te 63.3(1,2) 89.3(1)	C,C 96.2(2,4.0) C,S 90.7(2,4.4) 129.8(2,3.7) 164.8(2,8) C,Te 94.5(2,4.2) 165.0(2,4.7) S,Te 83.4(1,6)	365
[$Fe_3(\mu_3\text{-}S)(\mu_3\text{-}PPh)(CO)_9$][c] (dark red)	tr P-1 4	9.012(3) 15.151(4) 15.478(4)	108.20(4) 96.18(4) 87.39(4)	FeC_3PS (x3)	OC 1.799(7,37) μ_3p 2.208(1,44) μ_3S 2.264(2,23)	2.665(2,1) 3.499(1) S 72.33(7,0) 100.49(6) P 73.81(6,7) 106 33(5)	P,S 75.19(6,39)	371
				FeC_3PS (x3)	OC 1.784(4,52) μ_3p 2.210(1,45) μ_3S 2.260(1,17)	2.656(1,4) 3.511(1) S 72.11(3,17) 101.37(4) P 73.42(3,7) 106.75(4)	P,S 74.74(3,52)	
$Fe_3(\mu_3\text{-}S)(\mu_3\text{-}PBu^t)(CO)_8(CNR)$ (red) (at 213 K)	m $P2_1/c$ 4	8.709(3) 21.89(1) 13.614(5)	 127.73(2)	FeC_3SP (x2)	OC not given μ_3S 2.252(2,6) μ_3P 2.241(1,23)	2.666(2,24) S 103.3(0) P 105.5(0)	Fe,Fe 82.6(1) S,P 74.0(0,6)	368
				FeC_3SP (x1)	OC not given NC 1.971(4) μ_3S 2.239(1) μ_3P 2.201(1)		C,S 90.1(1) C,P 159.7(1) S,P 75.0(0)	

TABLE 6. Continued 25

$(OC)_9Fe_3(\mu_3\text{-}PBu^t)(\mu_3\text{-}Se)$ (red)	m $P2_1/c$ 4	9.078(5) 17.39(1) 14.479(9)	90.00(5)	FeC_3PSe (x3)	OC 1.770(17,43) μ_3P 2.206(4,1) 2.246(3) μ_3Se 2.637(2,4)	2.667(3,6) P 107.5(1) Se 97.6(1)	P,Se 76.6(1,6)	222
$(OC)_9Fe_3(\mu_3\text{-}S)(\mu_3\text{-}AsPh)$ (red) (at 233 K)	tr P-1 4	8.960(6) 15.20(1) 15.33(1)	109.14(6) 96.12(6) 87.31(6)	FeC_3SAs (x3)	OC 1.751 μ_3S 2.254(4,37) μ_3As 2.275(2,1) 2.359(2)	2.673(3,6) 3.550(3) S not given As not given	C,C 95.8(5,3.5) S,As 76.3(1,4)	370
$[(OC)_9Fe_3\{\mu_3\text{-}PMes\}_3]$.thf (red) (at 130 K)	or C2cm 4	8.358(3) 15.712(8) 25.184(9)		FeC_3P_2 (x3)	OC 1.791 μ_3P 2.234(3,13)	2.676(2,45) 3.566(2) P 106.2(1)	P,P 72.8(1,6) Fe,Fe 83.5(1)	232
$(OC)_9Fe_3(\mu\text{-}Cl)(\mu_3\text{-}SBu^t)$ (red)	m $P2_1/m$ 2	7.79(1) 15.22(3) 8.68(2)	107.8(1)	FeC_3ClS (x2)	OC 1.798 μCL 2.341(4) μ_3S 2.226(4)	2.681(4,0) 3.338(6) Cl 90.9(1) S not given	C,C 97.5(4,6.3)	372a
				FeC_3S (x1)	OC 1.798 μ_3S 2.127(4)		C,C 97.5(4,6.3)	
$(OC)_9Fe_3(\mu_3\text{-}SC_6H_{11})(\mu\text{-}SPh)$ (red orange) (at 233 K)	m $P2_1/c$ 4	13.31(2) 11.59(2) 26.27(3)	141.68(5)	FeC_3S_2 (x2)	OC 1.80(2,3) μ_3S 2.223(9,2) μS 2.331(8,2)	2.683(4,9) 3.419(5) μ_3S not given μS 94.3(2)	C,C 97.3(7,7.3) S,S 81.1(2,0)	215
				FeC_3S (x1)	OC 1.80(2,3) μ_3S 2.142(6)		C,C 97.3(7,7.3)	
$(OC)_9Fe_3(\mu_3\text{-}SC_6H_{11})(\mu\text{-}SePh)$ (red orange) (at 243 K)	m $P2_1/c$ 4	13.31(1) 11.679(9) 17.46(2)	110.82(8)	FeC_3SSe (x2)	OC 1.81(2,4) μ_3S 2.221(6,2) μSe 2.439(4,2)	2.690(4.13) 3.475(4) S not given Se 90.8(1)	C,C 96.9(9,6.6) S,Se 81.8(2,0)	215
				FeC_3S (x1)	OC 1.81(2,4) μ_3S 2.141(5)		C,C 96.9(9,6.6)	

TABLE 6. Continued 26

Compound	Cryst. cl. Space gr. Z	a, b, c	α, β, γ	Chromophore	Fe–L	Fe–Fe Fe–L–Fe	L–Fe–L	Ref.
$(OC)_6Fe_3(\mu_3\text{-}Se)(\mu_3\text{-}PBu^t)$. $(Ph_2PCCPh)_3$ (red) (at 220 K)	tr P-1 2	13.614(6) 14.121(5) 18.86(1)	109.53(4) 94.04(4) 106.36(4)	FeC_2P_2Se (x3)	OC 1.83(2) P 2.220(6,40) μ_3P 2.240(6,37) μ_3Se 2.381(2,15)	2.690(2,18) not given P 109.7(3) 99.6(1)	P,Se 74.8(2,5)	372b
$(OC)_9Fe_3(\mu_3\text{-}Se)(\mu_3Te)$ (dark ted)	tr P-1 2	6.9640(12) 9.360(2) 13.189(2)	94.090(16) 94.540(15) 110.670(16)	FeC_3SeTe (x3)	OC 1.794(6,17) μ_3Se 2.437(1,4) μ_3Te 2.480(1,17)	2.694(1,8) 3.635(1) Se 67.1(1,2) 96.6(1)	C,C 96.2(2,4.2) C.Se 91.1(1,6.1) 129.8(1,1.3) 166.4(1,2) C,Te 93.4(1,5.0) 166.1(2,2.5) SeTe 83.6(1,3)	365
$(OC)_9Fe_3(\mu_3\text{-}PPr^i)(\mu_3\text{-}Te)$ (brown)	m C2/c 8	30.02(1) 10.267(1) 13.867(1)	115.03(5)	FeC_3Te_2 (x3)	OC 1.805(16,35) μ_3P 2.194(3,5) 2.241(4) μ_3Te 2.551(3,11)	2.702(3,8) P 113.1(1) Te 91.6(1)	P,Te 77.1(1,5)	222
$(OC)_9Fe_3(\mu_3\text{-}SBu^t)(\mu\text{-}AsMe_2)$ (red orange) (at 221 K)	tr P-1 4	9.241(7) 16.23(1) 16.33(1)	110.55(6) 101.90(6) 91.61(6)	FeC_3SAs (x2)	OC 1.74(2,4) μ_3S 2.242(5,1) μAs 2.377(3,5)	2.706(3,7) 3.526(4) S not given As 96.1(1)	C,C 96.2(8,6.4) S,As 77.6(1,1)	215
				FeC_3S (x1)	OC 1.74(2,4) μ_3S 2.145(4)		C,C 96.2(8,6.4)	
$Fe_3(\mu_3\text{-}PPh)_2(CO)_6\{P(OMe)_3\}_3$ (carmone red)	tr P-1 2	10.539(1) 10.749(2) 18.762(4)	85.14(2) 75.50(1) 68.53(1)	FeP_3C_2 (x3)	P 2.142(1,9) μ_3P 2.233(1,20) OC not given	2.708(1,28) 3.622(1) P 108.9(1,6)	P,P 70.4(1,1) 103.0(1,4.2)	373
$(OC)_9Fe_3(\mu_3\text{-}PPh)_2$[c] (dark red)	tr P-1 4	11.446(8) 14.191(4) 15.586(7)	83.97(3) 86.47(5) 82.63(5)	FeC_3P_2 (x3)	OC 1.755(20,35) μ_3P 2.211(5,20)	2.715(3,2) P 106.6(2,3)	P,P 71.6(2,6)	374
				FeC_3P_2 (x3)	OC 1.75(2,4) μ_3P 2.214(5,8)	2.717(3,1) P 106.9(2,1)	P,P 71.5(2,2)	

TABLE 6. Continued 27

$(OC)_8Fe_3\{\mu_3\text{-}PPr^i\}$. $\{\mu\text{-Te-2,4,6-}Me_3C_6H_2\}$ (not given)	tr P-1 2	10.199(3) 10.587(4) 17.172(5)	81.05(3) 80.16(2) 71.11(2)	FeC_2Te_2P (x1)	OC not given μTe 2.544(1,5) μ_3P 2.235(2)	2.717(2,51) Te 64.5(1,1.6) P 126.6(1)	not given	222
				FeC_3PTe (x2)	OC not given μ_3P 2.228(2,1) μTe 2.533(1,6)		not given	
$[Fe_3(\mu_3\text{-}PH)_2(CO)_9]$ (red purple) (at 123 K)	tr P-1 2	9.159(6) 12.792(4) 6.975(2)	94.29(3) 112.03(3) 93.77(5)	FeC_3P_2 (x3)	OC not given μ_3P 2.216(2,11)	2.720(2,7) 3.561(2) P not given	not given	375
$Fe_3(\mu_3\text{-}PPh)_2(CO)_7(NCMe)_2$[c] (deep red)	or Pnma 8	13.103(2) 16.972(2) 25.038(5)		FeC_3P_2 (x2)	OC 1.786(9,19) μ_3P 2.229(1,42)	2.728(1,48) 3.554(1) P 75.8(1,1.7) 107.78(5)	C,C 98.7(4,5.2) P,P 69.7(1,4) C,P 98.1(2,10.0) 150.6(2,3.0)	376
				FeN_2P_2C (x1)	N 1.978(4,0) μ_3P 2.179(2,0) OC 1.731(7)		N,N 89.3(2) P,P 71.57(7) N,C 99.9(2) N,P 97.1(1) 161.0(1) P,C 96.6(2)	
				FeC_3P_2 (x2)	OC 1.787(9,37) μ_3P 2.229(2,14)	2.726(1,42) 3.549(1) P 75.8(1,2.0) 107.84(5)	C,C 99.1(4,4.2) P,P 69.7(1,5) C,P 97.6(2,11.7) 151.1(2,3.5)	
				FeN_2P_2C (x1)	N 1.967(4,0) μ_3P 2.177(1,0) OC 1.737(2)		N,N 87.8(2) P,P 71.68(7) N,C 101.6(2) N,P 96.8(1) 158.7(1) P,C 97.9(2)	

TABLE 6. Continued 28

$Fe_3(\mu_3\text{-}PC_6H_4Et)(PEt_3)_2(CO)_9$ (dark brown)	or $P2_12_12_1$ 4	11.723(6) 15.939(6) 16.442(7)		FeC_3P_2 (x2)	OC 1.77 μ_3P 2.219(4,6) P 2.149(5,1)	2.759(3,42)	not given	339
				FeC_3P (x1)	OC 1.77 μ_3P 2.144(4)			
$[Fe_3(\mu_3\text{-}Te)(PPh_3)_2(CO)_7]$ (dark red)	tr P-1 2	11.763(4) 13.598(5) 14.635(4)	87.66(3) 66.45(2) 14.635(4)	FeC_2Te_2P (x2)	OC 1.774(4,28) μ_3Te 2.541(1,10) P 2.244(1,30)	2.754(1,47) µTe 65.4(1,1.6) 98.6(1,3)	C,C 97.2(2,7) C,Te 91.9(2,7.5) 159.5(1,5.4) C,P 94.4(1,3.2) Te,Te 81.4(1,1) Te,P 96.3(1,2.3) 112.2(1,3.1) 172.4(1)	377
				FeC_3Te_2 (x1)	OC 1.777(5,9) μ_3Te 2.555(1,10)		C,C 97.3(2,3.1) C,Te 90.9(1,8.9) 129.6(2,5.6) 163.5(1) Te,Te 80.9(1)	
$(OC)_9Fe_3\{\mu\text{-}P(Ph)P(Ph)\text{-}1,2\text{-}C_6H_4PPh\}$ (red) (at 163 K)	m $P2_1/c$ 4	15.635(3) 11.856(2) 18.982(4)	112.85(2)	FeC_3P_2 (x1)	OC not given µP 2.255(1,39)	2.787(1,75) P 76.26(4) 109.54(5)	P,P 77.86(4) P,C 174.00(13)	378 379
				FeC_3P_2 (x1)	OC not given µP 2.2735(13) P 2.2362(13)			
				FeC_3P (x1)	OC not given µP	2.1787(14		

TABLE 6. Continued 29

$Fe_3(\mu\text{-}H)(\mu_3\text{-}PPr^i)(\mu\text{-}PMePr^i)(CO)_9$ (red)	tr P-1 2	8.715(2) 10.031(2) 14.481(2)	88.05(1) 80.37(1) 67.26(1)	FeC_3HP (x1)	OC not given μH 1.71(4) μ_3P 2.153(1)	2.790(1,7) 3.476(1) H 111(2) P 79.1(1,1) 100.5(1,19)	Fe,Fe 77.0(1)	355
				FeC_3HP_2 (x1)	OC not given μH 1.69(4) μ_3P 2.249(1) μP 2.294(1)		P,P 78.2(1)	
				FeC_3P_2 (x1)	OC not given μ_3P 2.211(1) μP 2.290(1)		P,P 79.0(1)	
$Fe_3(\mu\text{-}H)(\mu_3\text{-}PCH_2Ph)$. $\{\mu\text{-}PhCH_2P(CH_2Ph)_2\}(CO)_8$ (not given)	tr P-1 2	11.541(2) 17.739(2) 19.284(2)	78.01(1) 78.68(1) 90.36(1)	FeC_3HP_{2+} (x1)	OC 1.800(6,24) μH 1.64(5) μ_3P 2.249(1) μP 2.275(1)	2.795(1,32) 3.4975(8) H 111(3) μ_3P 80.1(-,1.5) 103.30(5) μP 101.05(5)	P,P 77.19(5) H,P 82(2,2)	380
				FeC_2HP_2 (x1)	OC 1.757(5,5) μH 1.71(5) μ_3P 2.112(1) P 2.228(1)		P,P 131.96(6) H,P 80(2,6) Fe,Fe 77.46(3)	
				FeC_3P_2 (x1)	OC 1.791(6,23) μ_3P 2.210(1) μP 2.256(1)		P,P 78.39(5)	
$(NEt_4)_2[Fe_3(\mu_3\text{-}Bi)_2(CO)_9]$ (dark brown)	or $P2_12_12_1$ 4	7.2800(10) 20.642(4) 22.365(4)		FeC_3Bi_2 (x3)	OC not given μ_3Bi 2.657(4,23)	2.804(5,10) Bi 63.8(1,4) 96.8(1,1)	Fe,Fe 90.65(14)	381

TABLE 6. Continued 30

C : Fe_3 isosceles triangle with one Fe - Fe bond

$Fe_3\{\mu-(cyhN)_2C\}(CO)_{10}$ (red)	m $P2_1/n$ 4	10.876(4) 16.609(7) 14.982(7)	90.7(1)	FeC_3N_2 (x2)	OC 1.81(2,3) μN 1.99(1,1)	2.409(3) 74.3(4,3)	C,C 94.2(9,6.6) N,N 64.6(5,2) C,N 102.6(8,5.6) 154.6(8,2.8)	382
				FeC_5 (x1)	OC 1.77(3,2) NC 2.00(2)		C,C 87(1,3) 116(1,12) 169(1)	
$[\{(OC)_3Fe\}_2(\mu-Cl)$. $\{\mu-P(Pr^i)CH_2P(Pr^i)Cl\}Fe(CO)_4]$ (red)	tr P-1 2	9.472(1) 12.414(2) 12.665(2)	63.55(1) 79.65(1) 80.88(1)	FeC_3ClP (x2)	OC not given μCl 2.316(1,2) μP 2.230(1,5)	2.5618(6) Cl 67.15(3) P 70.12(3)	Cl,P 81.5(1,1)	217
				FeC_4P (x1)	OC not given P 2.2119(8)			
$(ppn)[Fe_3(\mu-H)(\mu_3-PMe)$. $(\mu-CO)(CO)_{10}]$ (red orange)	m $P2_1/c$ 4	12.611(6) 22.644(8) 16.174(9)	99.86(7)	FeC_4HP (x2)	OC 1.76(1,3) μOC 1.95(1) μH not given μ_3P 2.235(4,10)	2.580(3) μ_3P 70.5(1) 125.3(2,1.6)	not given	383
				FeC_4P (x1)	OC 1.76(1,3) μ_3P 2.190(4)		not given	
$(OC)_{10}Fe_3\{\mu-\eta^3-P_3(Bu^t)_3\}$ (orange)	m $P2_1/n$ 4	9.180(5) 16.923(7) 19.186(9)	93.27(4)	FeC_3P_2 (x2)	OC 1.76(1,4) μP 69.9(1,1)	2.581(2) P 69.9(1,1)	C,C 94.8(6,5.0) P,P 68.2(1,2) C,P 93.3-142.2(4) 162.2(4,2.2)	237
				FeC_4P (x1)	OC 1.79(1,4) P 2.281(3)		C,C 87.0(6,1.0) 124.8(6) C,P 107.4(5,13.7)	

TABLE 6. Continued 31

$Fe_3(\mu_3\text{-}Te)(CO)_9)PPh_3)$	tr	9.891(3)	100.59(2)	FeC_3Te_2	OC 1.783(7,6)	2.585(1)	C,C 96.3(3,3.0)	384
(dark red)	P-1	12.140(3)	89.75(2)	(x2)	μ_3Te 2.571(1,6)	3.943(1,4)	Te,Te 75.2(2,4)	
	2	13.571(3)	105.49(2)			Te 60.35(2,1)	C,Te 98.4(2,10.9)	
						97.76(2,43)	155.4(2,4.0)	
				FeC_3Te_2P	OC 1.814(4,9)		C,C 90.5(2,1.0)	
				(x1)	μ_3Te 2.662(1,6)		173.4(2)	
					Ph_3P 2.297(1)		Te,Te 72.21(2)	
							C,Te 89.3(1,6.1)	
							167.5(1)	
							C,P 95.0(1,3.8)	
							Te,P 93.74(4)	
							165.88(3)	
$(PPh_4)[Fe_3(\mu_3\text{-}Te)_2(CO)_9I]$	tr	9.614(1)	74.47(1)	FeC_3Te_2	OC 1.782(11,11)	2.606(1)	C,C 96.4(4,3.8)	325
(not given)	P-1	14.511(2)	71.14(1)	(x2)	μ_3Te 2.574(1,22)	Te 60.9(1,2)	Te,Te 75.3(1,2)	
	2	14.668(2)	83.22(1)			97.1(1,8)	C,Te 97.6(3,8.4)	
							157.3(3,3.6)	
				FeC_3Te_2I	OC 1.802(6,27)		C,C 91.6(3,1.4)	
				(x1)	μ_3Te 2.632(1,18)		174.9(3)	
					I 2.666(1)		C,Te 92.9(3,4.4)	
							170.5(3)	
							C,I 90.4(3,5.1)	

TABLE 6. Continued 32

$(OC)_{10}Fe_3\{\mu\text{-}P(Ph)P(Ph)\text{-}1,2\text{-}$ $C_6H_4P(Ph)\}$[c] (red) (at 163 K)	tr P-1 4	10.499(1) 10.371(2) 34.046(7)	104.55(1) 89.43(1) 106.20(1)	FeC_3P_2 (x2)	OC not given μP 2.227(3,13)	2.613(2) P 71.8(1,2)	not given	378 379
				FeC_4P (x1)	OC not given P 2.242(2)		C,C 115.9(4) 177.8(4) C,P 89.8(3,2.6) 122.0(3,3.6)	
				FeC_3P_2 (x2)	OC not given μP 2.225(2,14	2.6032(13) P 71.6(1,3)	not given	
				FeC_4P (x1)	OC not given P 2.255(2)		C,C 103.0(4,6.2) 159.9(4) C,P 94.5(3,10.0) 146.1(3)	
$Fe_3(\mu\text{-}CO)\{\mu_3\text{-}\eta^3\text{-}Me_2PCH_2P.$ $(Me)CH_2PMe_2\}(CO)_{10}$ (orange)	m $P2_1/n$ 4	15.412(3) 9.346(2) 20.620(2)	110.42(1)	FeC_4P (x2)	OC 1.791(5,35) μOC 1.969(4,1) P 2.245(1,6)	2.705(1) C 86.7(2)	C,P 87.9(1,2.8) 173.3(1,3)	385
				FeC_4P (x1)	OC 1.783(5,6) P 2.245(1)		C,P 88.6(1,3.1) 173.6(1)	
$Fe_3)\{\mu_3\text{-}\eta^3\text{-}Et_2C(O)PPP(NEt_2)_2\}.$ $(CO)_{10}$ (orange)	m $P2_1/n$ 8	9.054(4) 38.75(3) 19.737(8)	104.05(3)	FeC_3P_2 (x2)	OC not given μP 2.315(14)	2.733(19)	not given	386
				FeC_4P (x1)	OC not given P not given		not given	
$(OC)_9Fe_3)\mu_3\text{-}PPh\}\{\mu\text{-}P(Ph)\text{-}1,2\text{-}$ $C_6H_4PPh\}$ (deep orange) (at 163 K)	tr P-1 2	9.885(2) 17.412(4) 9.799(2)	96.78(2) 95.46(2) 89.97(2)	FeC_3P_2 (x2)	OC not given μ_3P 2.209(1,2) μP 2.297(1,5)	2.8285(11) P 79.59(5) 103.4(1,2)	P,P 77.6(1,1) C,P 174.7(2,1.3)	379
				FeC_3P_3 (x1)	OC not given μ_3P 2.3196(13) μP 2.299(1,12)		P,P 78.14(5) C,P 165.7(2)	

TABLE 6. Continued 33

D : No Fe - Fe bond								
[$Fe_3(\mu$-$SPh)_6(CO)_6$].2MeCO[c] (dark red)	tr P-1 2	11.953(5) 13.88(1) 13.94(1)	89.51(6) 98.36(6) 99.12(0)	FeS_6 (x1)	μS 2.502(1,29)	3.170(1) S 81.6(1,6)	S,S 78.7(1,1.6)	387
				FeC_3S_3 (x2)	OC 1.800(1,11) μS 2.347(1,1)		C,C 95.1(1,1.0) S,S 85.1(4,1.3)	
				FeS_6 (x1)	μS 2.507(1,22)		S,S 78.0(1,9)	
				FeC_3S_3 (x2)	OC 1.804(4,12) μS 2.343(1,11)		C,C 95.1(1,1.0) S,S 84.6(1,3)	
$Fe_3(\mu$-$SPh)_6(CO)_6$ (black)	m $P2_1/n$ 2	11.077(1) 10.488(1) 19.052(2)	95.41(6)	FeS_6 (x1)	μS 2.504(1,18)	3.188(1) S 82.1(1,2)	S,S 78.3(1,1) 101.7(1,1) 180.00 Fe,Fe 180.00	388
				FeC_3S_3 (x2)	OC 1.804(4,6) μS 2.345(1,12)		C,C 95.2(2,1.0) S,S 84.8(1,5) C,S 89.9(1,4.7) 170.5(1,6)	
(ppn)[$Fe_3(\mu_3$-$As)(CO)_{12}$].thf[f] (red)	trg P-3 6	25.379(4) 22.442(4)		FeC_4As (x3)	OC not given μ_3As 2.462(2)	As 115.57(4)	not given	389 390
				FeC_4As (x3)	OC not given μ_3As 2.452(2)	As 115.72(4)	not given	
				FeC_4As (x3)	OC not given μ_3As 2.465(2)	As 115.47(4)	not given	
$(PPh_4)_2$[$Fe_3(\mu_3$-$As)(CO)_{12}H$] (red brown)	tr P-1 2	12.088(5) 13.257(5) 19.389(8)	80.43(3) 83.55(3) 70.83(3)	FeC_4As (x3)	OC 1.771(8,43) μ_3As 2.459(1,5)	4.160 μ_3As 115.6(1,7)	C,C 93.5(4,2.5) 117.4(2,4.7) C,As 87.6(2,4.7) 174.1(3,3)	391

TABLE 6. Continued 34

Compound	Crystal class, space group, Z	a, b, c	β	Chromophore	Fe–L	Fe–X–Fe	L–Fe–L	Ref.
$(PPh_4)_2[Fe_3(\mu_3\text{-}Sb)(CO)_{12}H]$ (black)	m $P2_1/c$ 4	11.165(5) 49.223(9) 10.903(5)	108.13(2)	FeC_4Sb (x3)	OC 1.761(8,46) μ_3Sb 2.596(1,2)	4.393 μ_3Sb 115.6(1,5)	C,C 93.9(4,2.2) 120.4(4,4.8) C,Sb 86.0(2,2.4) 117.7(3,4)	391
$(ppn)_2[Fe_3(\mu_3\text{-}SbCl)(CO)_{12}]$.thf (black)	m $P2_1/c$ 4	12.912(4) 27.567(4) 24.643(4)	92.16(2)	FeC_4Sb (x3)	OC not given μ_3Sb 2.55(-,1)	not given Sb 115(-,1)	not given	392
$(NEt_4)_2[Fe_3(\mu_3\text{-}SbCl)(CO)_{12}]$. $0.5CH_2Cl_2$ (deep red) (at 193 K)	m C2/c 8	29.30(1) 18.500(6) 16.844(6)	119.58(2)	FeC_4Sb (x3)	OC 1.770(9,40) μ_3Sb 2.566(2,4)	Sb 116.0(-,17)	C,C 93.0(4,6.0) 119.7(4,7.4) C,Sb 87.1(3,2.0) 175.0(3,3.8)	393
$(ppn)_2[(OC)_{10}Fe_3(\mu_3\text{-}Te)]$.0.5thf (red black) (at 223 K)	trg P-3 6	25.447(4) 22.432(4)		FeC_4Te (x3)	OC 1.76(2,4) μ_3Te 2.462(2,2)	not given Te 109.56(4,7)	C,C 92.9(7,2.7) 119.7(7,2.4) C,Te 87.1(4,3.3) 176.3(5,1.2)	319
$(NEt_4)_2[Fe_3(\mu_3\text{-}Sn)(CO)_{12}]^c$ (red)	m $P2_1/c$ 8	10.910(4) 40.45(1) 17.461(5)	98.00(2)	FeC_4Sn (x3)	OC not given μ_3Sn 2.548(3,9)	not given Sn 120.0(1,1.8)	not given	394
				FeC_4Sn (x3)	OC not given μ_3Sn 2.549(3,8)	not given Sn 120.0(1,1.2)	not given	
$(NEt_4)_2[Fe_3(\mu_3\text{-}Pb)(CO)_{12}]^c$ (black)	m $P2_1/c$ 8	10.959(2) 40.60(1) 17.455(4)	97.99(2)	FeC_4Pb (x3)	OC not given μ_3Pb 2.625(5,6)	not given Sn 120.0(1,2.0)	not given	394
				FeC_4Pb (x3)	OC not given μ_3Pb 2.623(5,1)	not given Sn 120.0(2,9)	not given	
$[(OC)_{10}Fe_3(\mu\text{-}Se_2)_2](SbF_6)_2.SO_2$ (orange brown)	m $P2_1/c$ 4	11.810(9) 24.023(6) 10.853(7)	107.15(5)	FeC_3Se_2 (x2)	OC 1.86(1,1) μSe 2.406(3,31)	not given Se 93.49(9)	C,C 92.0(6) Se,Se 57.8(1,1) 91.4(1,5.6)	395
				FeC_4Se_2 (x1)	OC 1.875(20,5) μSe 2.415(3,0)		C,C 92.9(6) Se,Se 93.06(9)	

TABLE 6. Continued 35

$(NEt_4)_2[Fe_3\{\mu-\eta^2-P_2(H_2)$. $(N_2Pr^i_2)\}(CO)_{12}]$ (red brown)	m $P2_1/n$ 4	10.018(5) 29.892(13) 12.337(5)	100.66(4)	FeC_4P (x3)	OC 1.76(2,7) µP 2.298(4,48)	not given P 120.9(2)	not given	396
$Fe_3\{\mu-Et_2NCOPPP(NEt_2)_2\}$. $(CO)_{12}$ (deep orange)	m $P2_1/n$ 4	17.071(7) 19.117(8) 10.853(3)	96.82(3)	FeC_4P (x2)	OC not given P 2.247(4,24)	not given	not given	386
				FeC_4P_2 (x1)	OC not given C 2.00(1) µP 2.302(3) P 2.312(4)		P,P 78.6(1) C,P 67.7(3)	
$Fe_3\{\mu_3-\eta^3-Me_2PCH_2P(Me)$. $CH_2PMe_2\}(CO)_{12}$ (orange)	tr P-1 2	10.416(2) 12.007(2) 13.010(3)	100.373(8) 103.244(9) 111.29(1)	FeC_4P (x3)	OC 1.785(3,7) P 2.231(1,5)		C,C 91.0(1,1.5) 119.9(1,6.3) C,P 89.1(1,6.7) 176.0(1,2.9)	385

a Where more than one chemically equivalent distance or angle is present, the mean value is tabulated. The first number in parenthesis is e.s.d., the second is maximum deviation from the mean value.
b The chemical identity of coordinated atom/ligand is specified in these columns.
c There are two crystallographically independent molecules.
d Five - membered metallocyclic ring.
e Three - membered metallocyclic ring.
f There are three crystallographically independent molecules.

Table 7. Crystallographic and Structural Data for Tetranuclear Iron Carbonyls[a]

COMPOUND (COLOUR)	CRYST.CL. SPACE GR. Z	a[Å] b[Å] c[Å]	α[°] β[°] γ[°]	CHROMO-PHORE	Fe–L [Å]	Fe–Fe [Å] Fe-L-Fe [°]	L–Fe–L [°]	REF.
A : Planar (tetrahedral) array of Fe_4								
$Fe_4(\mu_4\text{-NEt})_2(CO)_{11}$ (black)	m C2/c 8	13.777(2) 14.014(3) 21.294(4)	 91.45(3)	FeC_3N_2 (x2)	OC[b] not given μ_4N 2.018(5,20)	2.406(1,15) N not given	C,C[b] not given	397
				FeC_4N_2 (x2)	OC not given μ_4N 2.050(5,27)			
$Fe_4(\mu_4\text{-NEt})(\mu_7\text{-}\eta^2\text{-ONEt})$. $(\mu\text{-CO})_3(CO)_8$ (deep green)	tr P-1 2	14.079(5) 9.461(4) 8.770(4)	112.74(4) 91.62(1) 92.05(1)	FeC_3ON (x1)	OC not given μO 1.95(1) μ_4N 2.07(1)	2.468(3,2) 2.541(3,11) not given	Fe,Fe 90.0(1,1.1)	398
				FeC_4ON (x1)	OC not given μO 1.94(1) μ_4N 2.02(1)			
				FeC_3N_2 (x1)	OC not given μ_4N 1.99(1,3)			
				FeC_4N_2 (x1)	OC not given μ_4N 1.99(1,7)			
$[Fe(py)_6][Fe_4(\mu_3\text{-CO})(\mu\text{-CO})_3$. $(CO)_2]$ (black)	tr P-1 2	10.09(2) 14.86(4) 15.61(4)	90.00(10) 90.00(10) 103.12(10)	FeC_4 (x3)	OC 1.69(4,5) μ_3OC 2.00(3,2) μOC 1.82(4,3)	2.50(x3) 2.58(-,2)	C,C 93.7(1.6,8.9) 167.5(1.4,1.5)	399
				FeC_3 (x1)	OC 1.78(3,3)		C,C 98.2(1.4,3)	
				FeN_6 (x1)	pyN 2.26(3,4)		N,N 90.0(1.5,2.0) 179.1(1.5,7)	

Table 7. Continued 2

$Fe_4(\mu_4\text{-}S)_2(\mu\text{-}CO)_3(CO)_8$ (dark green)	or Pccn 4	6.603(1) 15.429(3) 17.292(4)		FeC_4S_2 (x2)	OC 1.78(1,1) μOC 1.93(1) μ_4S 2.339(2,8)	2.510(3,22) 2.605(2) C 80.4(6) S 66.6(-,2.8) 101.8(-,5)	C,C 91.1(4,9) S,S 77.0(1) C,S 95.6(2,1.9) 170.5(3,1.5)	400
				FeC_3S_2 (x2)	OC 1.76(1,1) μOC 1.79(1) μ_4S 2.286(2,8)		C,C 93.3(4,4.2) S,S 79.1(1) C,S 98.8(3,7.9) 158.9(3,7)	
(ppn)$[Fe_4(\mu_3\text{-}CO)(CO)_{12}]$ (dark red) (at 115 K)	rh R-3 3	18.039(7)	81.51(3)	FeC_4 (x3)	OC 1.780(12,33) μ_3OC 2.027(24,20)	2.550(5,6) 2.601(6,9) C 78.0(11,1.1)	C,C 96.4(13,13.0) 155.1(6,3)	401
				FeC_3 (x1)	OC 1.801(12,10)		C,C 97.9(7,7)	
$Fe_4(\mu_3\text{-}CO)_2(CO)_{10}$ (brown)	or Pnma 4	12.811(4) 13.216(2) 11.788(2)		FeC_4	OC not given μ_3OC 2.070(4,38)	2.529(1,5) 2.272(1) not given	not given	402
$Fe_4(\mu_4\text{-}S)(_4\text{-}PPh)(\mu\text{-}CO)(CN).$ $(CO)_8$ (dark green) (at 213 K)	tr P-1 2	8.716(4) 9.757(4) 14.068(5)	91.29(3) 93.05(4) 103.16(3)	FeC_3SP (x2)	OC 1.739-1.790 μOC 1.895(4,11) μ_4S 2.300(1,2) μ_4P 2.262(1,10)	2.503(1,1) 2.685(1,7) C 79.9(1,1) P 107.6(0,2.6)	Fe,Fe 83.9(0) 90.0(0,2.4) S,P 75.2(0,3)	403
				FeC_4SP (x1)	OC 1.739-1.790 NC 1.952(4) μOC 2.002(3,15) μ_4S 2.478(1) μ_4P 2.335(1)		S,P 70.5(0) C,S 99.2(1) C,P 169.7(1)	
				FeC_3SP (x1)	OC 1.739-1.790 μ_4S 2.291(1) μ_4P 2.227(1)		S,P 76.0(0)	

Table 7. Continued 3

Compound (colour)	Crystal class, space group, Z	a, b, c	α, β, γ	Chromophore	Fe-L	Fe-Fe, Fe-L-Fe	L-Fe-L	Ref.
$\{Fe_2(CO)_4(py)_2\}_2$[c] (dark brown)	m $P2_1/a$ 8	21.511(3) 16.766(3) 19.253(3)	 117.94(2)	FeC_4 (x2)	OC 1.725(25,49)	2.543(3,4) 2.757(3)	C,C 104.7(10,2) 149.0(8,5)	404
				FeN_2 (x2)	pyN 2.144(12,8)		N,N 88.0(4,1.3)	
				FeC_4 (x2)	OC 1.735(25,24)	2.544(3,6) 2.761(3)	C,C 104.8(10,1.2) 147.6(9,4)	
				FeN_2 (x2)	pyN 2.174(12,11)		N,N 88.4(4,3)	
$Fe_4(\mu_4\text{-}Se)_2(\mu\text{-}CO)(CO)_{10}$ (not given)	not given	not given		FeC_3Se_2 (x2)	OC not given μOC not given μ_4Se 2.453(2,6)	2.535(2) 2.594(2,1) 2.694(2)	not given	405
				FeC_3Se_2 (x2)	OC not given μ_4Se 2.401(2,6)		not given	
$Fe_4(\mu_4\text{-}Ptol\text{-}p)_2(\mu\text{-}CO)(CO)_9$. $\{P(OMe)_3\}$ (black)	or $P2_12_12_1$ 4	12.890(8) 14.785(4) 18.197(3)		FeC_3P_2 (x3)	OC 1.79(2,5) μOC not given μ_4P 2.324(5,24) 2.259(5,6)	2.458(3) 2.656(3,15) 2.755(3) P 108.8(2,1.2)	P,P 70.8(2,1.4)	406
				FeC_2P_3 (x1)	OC 1.79(2,5) μ_4P 2.244(5,12) P 2.170(5)		not given	
$Fe_4(\mu_4\text{-}Ptol\text{-}p)_2(CO)_{11}\{P(OMe)_3\}$ (red)	tr P-1 2	13.000(2) 15.069(2) 9.445(2)	100.65(1) 107.35(1) 84.43(1)	FeC_3P_2 (x3)	OC 1.79(1,5) μ_4P 2.221(2,13) 2.379(2,27)	2.667(2,1) 2.698(2,12) P 111.1(1,6.1)	P,P 68.8(1,5)	406
				FeC_2P_3 (x1)	OC 1.79(1,5) μ_4P 2.197(2) 2.351(2) P 2.184(2)		not given	

Table 7. Continued 4

$[Fe_4\{\mu_8\text{-}\eta^3\text{-PhPPPPh}\}.$ $(\mu\text{-}PPh_2)(CO)_{10}].0.5C_5H_{12}$ (black)	m $2_1/c$ 4	11.926(1) 16.631(2) 21.900(3)	103.09(2)	FeC_2P_3 (x1)	OC not given μPh_2P 2.196(2) μP 2.205(2) μ_4P 2.266(2)	2.725(1,1) 2.792(1,32) P 76.7(-,1) 119.8(-,4) μ_4P 71.95(6)	Fe,Fe 90.0(-,1.5)	407
				FeC_2P_3 (x1)	μPh_2P 2.189(2) μP 2.297(2) μ_4P 2.250(2)			
				FeC_3P_2 (x1)	μP 2.342(2) μ_4P 2.241(2)			
				FeC_3P_2 (x1)	μPh_2P 2.247(2) μ_4P 2.261(2)			
$Fe_4(\mu_4\text{-}PPh)_2(\mu\text{-}PPh)_2(CO)_8$ (red)	tr P-1 1	10.891(4) 11.655(5) 9.766(4)	114.17(1) 99.45(1) 68.40(1)	FeC_3P_2	not given	not given	not given	408
$Fe_4(\mu_4\text{-}\eta^2\text{-NCPh})(CO)_{12}$ (not given)	m $P2_1/c$ 4	9.492(2) 17.641(4) 14.170(3)	98.34(2)	FeC_3N (x2)	not given	not given	not given	409
				FeC_4N (x2)	not given			
B : Butterfly								
$(ppn)[Fe_4(\mu_4\text{-}N)(CO)_{12}]$ (black)	tr P-1 2	10.700(2) 14.146(3) 16.249(3)	91.15(1) 94.55(2) 97.52(2)	FeC_3N	OC 1.788(9,28) μ_4N 1.778(5,4) 1.900(5,4)	2.512(1) 2.604(1,8) 3.541(1) N 88.8(2,6.0) 179.0(2)	C,C 98.1(4,3.6) 104.5(3,2.4) 124.7(3,6) 148.7(3)	410

Table 7. Continued 5

(ppn)[$Fe_4(\mu_4$-N)$(CO)_{11}(PPh_3)$] (brown)	tr P-1 2	11.009(6) 17.285(4) 17.380(2)	103.11(3) 91.18(2) 105.26(3)	FeC_3N (x3)	OC 1.772(9,27) μ_4N 1.778(5) 1.900(5,22)	2.507(1) 2.592(1,3) 2.651(1) N 88.7(2,4.4) 176.0(3)	C,C 97.5(4,3.1) C,N 105.5(3,1.9) 126.2(2,5.0) 143.8(3)	411
				FeC_2NP (x1)	OC 1.771(6,35) μ_4N 1.771(5) Ph_3P 2.217(2)		C,C 97.9(3) C,N 103.2(2) 148.9(3) C,P 93.4(2,5) N,P 108.0(2)	
$Fe_4(\mu$-H)$(\mu_4$-N)$(CO)_{12}$ (brown)	tr P-1 2	7.491(2) 9.214(3) 13.974(6)	88.09(2) 86.98(3) 73.76(3)	FeC_3HN (x2)	OC not given μH not given μ_4N 1.84(2,8)	2.58(2,4) N 178.4(6)	not given	412
				FeC_3N (x2)	OC not given μ_4N not given			
$Fe_4(\mu_4$-PPr$^i)(CO)_{12}$ (not given)	m $P2_1/c$ 4	9.555(5) 22.31(1) 13.925(8)	 133.78(3)	FeC_3P	OC 1.786 μ_4P 2.209(3,5) 2.275(2,1)	2.564(2,13) 2.709(2) P 75.6(1) 127.9(1)	Fe,Fe 105.8(1,1)	413
$Fe_4(\mu$-H)$(CO)_{12}(\mu$- η^2-COMe) (not given)	tr P-1 -	9.137(2) 9.313(2) 14.759(4)	97.13(2) 99.32(2) 124.15(2)	FeC_4H (x2)	OC 1.790(2) μOC 1.960(2,13) μH 1.69	2.621(1,24) 3.925(1)	not given	414
				FeC_4O (x1)	OC 1.790(2) μOC 2.138(2) O 2.000(2)		not given	
				FeC_4 (x1)	OC 1.790(2) μOC 1.834(2)		not given	
$HFe_4(\mu_6$-η^3-$BH_2)(CO)_{12}$ (brown)	m $P2_1/c$ 4	16.429(5) 8.740(3) 13.237(3)	 94.32(1)	FeC_3HB	OC 1.790(7,14) μH 1.62(5,10) μ_4B 2.008(6,42)	2.662(1,25) H 104.2(23) B 82.5(2,2.2)	C,C 94.9(3,4.7) H,B 43.8(19,4) 89.1(15)	415

Table 7. Continued 6

Compound (colour)	Cryst. cl. / Space group / Z	a, b, c	α, β, γ	Chromophore	Fe–L	Fe–Fe / Fe–L–Fe	L–Fe–L	Ref.
[Fe_4(μ-H)($μ_4$-PH){μ-(PPr^i)$_2$$CH_2$}. $(CO)_{10}$] (not given)	m $P2_1/n$ 2	9.789(7) 24.661(3) 14.906(2)	 97.72(1)	FeC_3P_2 (x2)	OC 1.780(6,13) $μ_4$P 2.286(1,2) μP 2.223(1,2)	2.654(1,3) 2.790-(1) 4.109(1) $μ_4$P 69.3(-,1.5) 131.0(-,3.0) P 75.5(-,4.7)	Fe,Fe 104.4(-,2)	416
				FeC_2P_3 (x2)	OC 1.766(6,35) $μ_4$P 2.407(1,3) 2.166(2,3) 2.214(2,2)		not given	
[{Fe_2(μ-SPh)$(CO)_2$}$_2$($μ_4$-S)] (dark red)	tr P-1 2	15.389(6) 8.969(4) 12.464(5)	75.43(2) 99.42(2) 103.22(2)	FeC_3S_2	OC 1.757(14,21) μPhS 2.266(3,7) $μ_4$S 2.241(3,5)	2.521(2,2) μS 68.8(1,1) 132.8(1,6.5)	C,C 96.3(6,6.9) S,S 76.7(1,4) C,S 97.6(4,8.7) 157.1(5,4.7)	417
[$(OC)_6$$Fe_2$(μ-SEt)($μ_4$-S). Fe_2(μ-SPh)$(CO)_6$] (red)	m $P2_1$ 4	9.111(2) 12.130(2) 25.886(5)	 90.52(2)	FeC_3S_2	OC not given μS 2.251(4) $μ_4$S 2.232(4)	2.522(3,4) $μ_4$S 68.4(1) 126.2(2)	S,S 76.1(1)	418
[{Fe_2(μ-SEt)$(CO)_6$}$_2$($μ_4$-S)] (dark red)	or $P2_12_12_1$ 4	11.875(2) 12.920(3) 17.020(3)		FeC_3S_2	OC 1.813(11,53) μEtS 2.267(3,8) $μ_4$S 2.246(2,8)	2.540(2,2) μS 68.14(8,7) $μ_4$S 68.9(-,2) 132.9(1,5.8)	C,C 95.8(5,6.7) S,S 77.3(1,5) C,S 97.6(4,8.8) 157.7(4,2.0)	419
[{Fe_2(μ-SMe)$(CO)_6$}$_2$($μ_4$-S)] (deep red)	tr P-1 2	8.93(2) 9.28(2) 16.04(3)	81.67(17) 80.08(17) 66.58(17)	FeC_3S_2	OC 1.74(2,4) μMeS 2.247(8,10) $μ_4$S 2.248(7,9)	2.540(7,4) 4.119(11,85) μS 67.9(2,4) $μ_4$S 68.8(2,1) 132.8(8,5.0)	C,C 95.6(10,7.5) S,S 77.1(3,1) C,S 101.6(9,11.4) 157.6(8,8.0)	420
[{Fe_2(μ-SPh)$(CO)_6$}$_2$($μ_4$Se)] (red)	m $P2_1/n$ 4	8.971(3) 12.891(8) 26.955(7)	 98.03(2)	FeC_3SSe	OC not given μPhS 2.270(5,6) $μ_4$Se 2.347(3,8)	2.553(3,0) μS not given $μ_4$Se 65.9(9,3) 134.8(1,7.0)	S,Se 77.1(1,9)	421

Table 7. Continued 7

[{Fe_2(μ-SEt)}$_2$(CO)$_6$(μ$_4$-Se)] (dark red)	or $P2_12_12_1$ 4	12.103(2) 12.884(3) 17.083(3)		FeC_3SSe	OC 1.79(2,5) μEtS 2.269(6,19) μ$_4$Se 2.357(3,2)	2.571(4,5)	C,S 103.5(7) 157.8(6,1.8) C,Se 91.4(6) 157.3(8,1.4) S,Se 77.5(2,5)	214
[{Fe_2(μ-Setol)(CO)$_6$}$_2$(μ$_4$-Se)] (dark red)	m $P2_1/n$ 8	11.059(2) 32.367(6) 18.933(3)	97.14(3)	FeC_3Se_2	OC not given μSe 2.392(4,8) μ$_4$Se 2.353(3,6)	2.601(4) μSe 65.5(1,4) μ$_4$Se 66.8(1,4) 133.9(1,7.5)	Se,Se 78.5(1,6)	422
[{Fe_2(μ-SeBu)(CO)$_6$}$_2$(μ$_4$-Se)] (orange)	or Pbca 16	16.014(3) 19.137(4) 39.244(9)		FeC_3Se_2	OC not given μSe 2.410(11,24) μ$_4$Se 2.354(11,32)	2.617(12,1) μSe 65.8(3,4) μ$_4$Se 67.5(3,2) 137.2(4,14)	not given	423
[(OC)$_8$$Fe_2$(μ$_4$-P)$Fe_2(CO)_6$(μ-Cl)] (red)	m $P2_1/c$ 4	10.28(1) 18.19(3) 16.08(3)	133.20(8)	FeC_4P (x2)	OC 1.794(2,2) μ$_4$P 2.28	2.540(2) 2.733(1) P 71.3(1,2.8) 131.3(1,2.2) Cl 66.9(1)	not given	424
				FeC_3ClP (x2)	OC 1.793(2,5) μCl 2.30 μ$_4$P 2.245(-,10)		not given	
Fe_4(μ$_4$-P){μ-PhCH$_2$PCH$_2$. P(CH$_2$Ph)$_2$}(CO)$_{10}$ (not given)	m C2/c 8	36.357(2) 10.752(2) 22.879(5)	112.45(1)	FeC_3P_2 (x2)	OC 1.784(7,42) μ$_4$P 2.247(2,1) μP 2.225(2,11)	2.598(1) 2.723(1) μ$_4$P 72.0(-,1.5) 130.1(-,4.3) P 71.46(5)	P,P 77.0(-,2)	380
				FeC_4P (x1)	OC 1.811(7,21) μ$_4$P 2.292(2)		not given	
				FeC_3P_2 (x1)	OC 1.796(7,14) μ$_4$P 2.262(2)		P,P 96.50(6)	
[{Fe_2(CO)$_8$}$_8$(μ$_4$-Si)] (orange)	m $P2_1/c$ 4	9.655(4) 11.933(4) 20.695(5)	98.47(3)	FeC_4Si	OC 1.80(1,5) μ$_4$Si 2.347(4,5)	2.792(3,1) Si 73.1(1,1)	not given	425

Table 7. Continued 8

$[Fe_4\{\mu_6\text{-}\eta^3\text{-}PP(Ph)P(Ph)\}.$ $(\mu\text{-}PPh_2)(CO)_{12}]^c$ (yellow)	or $Pna2_1$ 8	20.781(2) 37.222(3) 10.378(1)		FeC_3P_2 (x2)	OC not given μP 2.208(2,37)	2.629(1) 2.6919(9) P 73.8(-,1.4)	not given	407
				FeC_3P_2 (x2)	OC not given μP 2.221(1,11) μPh_2P 2.228(1,4)			
				FeC_3P_2 (x2)	OC not given μP 2.210(1,35)	2.619(1) 2.655(1) P 73.5(-,2.1)	not given	
				FeC_3P_2 (x2)	OC not given μP 2.228(1,24) μPh_2P 2.166(1,5)			
D ; Two $XFe_2(CO)_6$ fragments joined by > X - X < bond :								
$(AsPh_4)_2[\{Fe_2(\mu\text{-}S)(CO)_6\}_2.$ $(\mu_4\text{-}\eta^2\text{-}S_2)]$ (deep orange)	m $P2_1/n$ 2	10.662(3) 18.397(4) 16.245(7)	107.66(4)	FeC_3S_2	OC not given μS 2.291(1,6) μ_4S_2 2.245(1,0)	2.518(1) S 66.7(1) S_2 68.2(1)	S,S 75.3(1,1)	426
$[\{Fe_2(\mu\text{-}SPh)(CO)_6\}_2(\mu_4\text{-}\eta^2\text{-}S_2]^c$ (red orange)	tr P-1 2	10.584(4) 11.247(1) 14.257(3)	104.20(2) 90.80(4) 98.12(2)	FeC_3S_2	OC 1.798(10,33) μ_4S_2 2.246(2,8) μPhS 2.277(2,5)	2.523(1,3) S 67.8(1,6)	C,C 96.2(4,5.7) S,S 78.5(1,3) C,S 97.2(3,12.6) 156.9(3,4.1)	427
$(PPh_s)_2[\{Fe_2(\mu\text{-}Se)(CO)_6\}_2.$ $(\mu_4\text{-}\eta^2\text{-}Se_2)]$ (black)	m $P2_1/n$ 4	10.561(6) 14.021(5) 20.46(1)	91.89(4)	FeC_3Se_2	OC 1.80(1) μSe 2.416(5,1) μ_4Se_2 2.356(2)	2.555(6) Se 64.7(2,1.0) 108.1(2,1.1)	C,C 98(1,4) Se,Se 73.2(2,1) C,Se 84-161(1)	428
$(HNEt_3)[\{Fe_2(\mu\text{-}SBu^n)(CO)_6\}_2.$ $\{\mu_4\text{-}\eta^2\text{-}As(Me)AsMe\}]$	m $P2_1/c$ 4	14.779(3) 14.780(2) 17.215(6)	114.25(2)	FeC_3SAs	OC 1.796(9,55) μS 2.268(2,4) μAs 2.339(2,3)	2.591(1,5) S 69.7(-,1) As 67.3(-,2)	C,.C 95.4(4,7.0) C,S 101.0(4,4.9) 156.4(3,2.5) C,As 97.2(3,13.5) 153.9(3,4.7) S,As 75.1(-,8)	429

Table 7. Continued 9

E : Unique structures								
$Fe_4(\mu_8\text{-}\eta^4\text{-}S_4C_2)(CO)_{12}$ (orange red)	tr P-1 1	6.541(2) 7.744(5) 12.619(6)	102.17(2) 102.07(2) 103.20(2)	FeC_3S_2	OC not given µS 2.293(2,5)	2.486(2) S 65.7(1,1)	S,S 73.5(1,1)	430
$[\{Fe_2(\mu\text{-}SMe)(CO)_6\}_2\cdot$ $\{\mu_4\text{-}\eta^2\text{-}SCH_2C_6H_4CH_2S\text{-}m\}]$ (red)	tr P-1 2	7.831(2) 14.423(3) 14.928(2)	68.29(1) 87.76(2) 83.19(2)	FeC_3S_2	OC 1.786(9,34) µMeS 2.257(2,6) µLS 2.263(2,12)	2.514(1) S 67.5(-,2)	C,C 96.5(4,6.9) S,S 80.8(-,4) C,S 96.7(3,8.2) 157.1(3,5.7)	431
$[Fe_3(\mu\text{-}CO)(\mu\text{-}S)(CO)_6\cdot$ $Fe(\mu\text{-}S_2C_6H_{11})(CO)_3]$ (red)	m $P2_1/c$ 4	14.825(7) 10.707(6) 16.801(8)	 106.56(3)	FeC_2S_3 (x1)	µOC 1.815 µS 2.215(2) µLS 2.256(2,0)	2.559(1,13) 2.639(1,2) not given	C,S 87.3(-,2.2) 145.0 Fe,Fe 58.3 149.7(-,1.8)	432
				FeC_4S (x2)	OC not given µS not given			
				FeC_4 (x1)	OC not given µOC not given			
$[Fe_2(\mu\text{-}\eta^2\text{-}B_2H_4)(CO)_6]_2$ (dark red) (at 113 K)	m $P2_1$ 2	7.3740(12) 13.956(3) 10.627(2)	 106.22(2)	$FeC_3H_2B_2$	OC not given µH 1.51(6,3) µB 2.153(10,65)	2.578(2) not given	not given	433
$[Fe_2(\mu\text{-}\eta^2\text{-}B_2H_4)(CO)_6]_2$ (red orange) (at 123 K)	or Pbca 8	12.990(4) 12.279(3) 26.527(10)		$FeC_3H_2B_2$	OC not given µH 1.61(3,8) µB 2.139(2,66)	2.591(1) not given	not given	433
$(PPh_4)[Fe_4(\mu_4\text{-}P)(\mu\text{-}CO)(CO)_{13}]$ (red)	m $P2_1/c$ 2	11.467(5) 28.590(4) 12.600(6)	 100.80(4)	FeC_4P (x2)	OC not given µOC not given μ_4P 2.241(2,4)	2.593(2) 2.630(2,11) not given	not given	434
				FeC_4P (x1)	OC not given μ_4P 2.234(2)			
				FeC_3P (x1)	OC not given μ_4P 2.156(2)			

Table 7. Continued 10

Compound	Crystal system, space group, Z	a, b, c	β	Chromophore	Fe–L	Fe–Fe, Fe–X–Fe	L–Fe–L	Ref.
$(PPh_4)_2[Fe_4(\mu\text{-}H)(\mu_4\text{-}P)(CO)_{13}]$ (red)	tr P-1 2	10.968(2) 11.886(2) 25.397(4)	97.84(2) 95.86(1) 113.59(1)	FeC_3P (x3)	OC 1.77(1,4) μ_4P 2.187(3,17)	2.653(2,1) 2.696(2) not given	not given	435
				FeC_4P (x1)	OC 1.78(1,2) μ_4P 2.275(3)		not given	
$(NEt_4)[Fe_4(\mu_4\text{-}Sb)(\mu\text{-}CO)(CO)_{13}]$ (black)	m $P2_1/n$ 4	15.053(5) 11.051(4) 19.979(6)	110.86(2)	FeC_4Sb (x2)	OC 1.77(2,5) μOC 2.00(2,1) μ_4Sb 2.516(3,2)	2.660(3) 2.682(4,3) Sb 64.8(1,1.0) 141.4(1,7.0)	Fe,Fe 60.0(1,6) Fe,Sb 57.6(1,1.3)	436
				FeC_3Sb (x1)	OC 1.78(2,1) μ_4Sb 2.460(3)			
				FeC_4Sb (x1)	OC 1.76(2,7) μ_4Sb 2.481(3)			
$(NEt_4)_2[Fe_4(\mu\text{-}H)(\mu_4\text{-}Sb)(CO)_{13}]$ (brown)	m P_2/c 4	13.805(6) 18.934(9) 15.295(6)	97.76(3)	FeC_3HSb (x2)	OC 1.75(1,3) μH not given μ_4Sb 2.470(2,4)	2.718(2,4) 2.764(2) Sb 66.9(-,2) 138.1(-,1.6)	C,C 100.3(6,7.0) C,Sb 97.0(3,5.7) 147.6(3,2.5)	393
				FeC_3Sb (x1)	OC 1.76(1,1) μ_4Sb 2.460(2)		C,C 100.1(6,1.7) C,Sb 99.1(3,7) 149.1(3)	
				FeC_4Sb (x1)	OC 1.75(1,2) μ_4Sb 2.521(2)		C,C 93.7(6,2.9) 119.6(5,5.0) C,Sb 86.4(4,1.5)	

Table 7. Continued 11

$(NEt_4)[Fe_4(\mu\text{-}H)_2(\mu_4\text{-}Sb)(CO)_{13}]$ (black)	tr P-1 2	12.714(2) 14.342(2) 8.970(2)	99.57(1) 105.27(1) 82.34(1)	FeC_3HSb (x2)	OC 1.778(5,18) μH 1.64(5,4) μ_4Sb 2.464(-,6)	2.7340(9) 2.779(1,6) Sb 67.8(-,4) 139.9(-,2.4)	C,C 98.4(2,5.3) C,H 87(2,15) 167(2,5) C,Sb 97.3(2,6.5) 151.1(2,7) H.Sb 90(2,1)	393
				FeC_3H_2Sb (x1)	OC 1.794(5,11) μH 1.68(5,4) μ_4Sb 2.489(-,19)		C,C 97.3(2,2.4) H,H 70(2) C,H 88.5(2,15.5) 166.5(2,6.5) C,Sb 92.6(1,2.4) 163.2(2) H.Sb 88(2,1.0)	
				FeC_4Sb (x1)	OC 1.785(5,11) μ_4Sb 2.483(8,24)		C,C 92.9(2,9) 120.1(2,1.4) C,Sb 87.1(1,6) 178.9(2)	
$Fe_4(\mu\text{-}CO)\{\mu_3\text{-}\eta^3\text{-}Me_2PCH_2P.$ $(Me)CH_2PMe_2\}(CO)_{13}$ (green)	m $P2_1/n$ 4	12.135(2) 11.757(2) 24.700(2)	 104.202(8)	FeC_4P	OC not given P 2.222(7,38)	2.595(5)	not given	385
$Fe_4\{\mu_4\text{-}\eta^2\text{-}P(Me)PMe\}_2(CO)_{12}$ (red)	m C2/c 4	10.856(1) 16.245(2) 15.000(7)	 102.25(3)	FeC_3P_2	OC 1.795(6,15) μP 2.212(2,3) 2.236(2,1)	2.618(1) P 72.0(1,1)	P,P 77.8(1,1) P,Fe 54.0(1,5)	436
$(PPh_4)_2[Fe_4(\mu_3\text{-}Te)_2(CO)_{14}]$ (dark brown)	m C2/c 4	23.324(2) 11.793(2) 23.499(5)	 102.02(1)	FeC_3Te_2 (x2)	OC not given μ_3Te 2.576(3,0)	2.641(5) not given	Te,Te 77.48(8)	348
				FeC_4Te (x2)	OC not given μ_3Te 2.632(2,0)			
$(NMe_4)[Fe_4(\mu_4\text{-}Sb)(CO)_{16}]$ (red)	tr P-1 2	11.840(3) 12.550(3) 10.500(2)	94.41(2) 101.14(2) 89.20(2)	FeC_4Sb	OC 1.793(8,42) μ_4Sb 2.569(1,3) 2.637(1,6)	2.720(1) Sb 62.09(2) 117.1(-,4.0)	C,C 92.8(3,8.3) 119.9(3,2.6) 172.4(2,4.8) C,Sb 89.5(2,6.8) 171.6(2,5.6)	437

Table 7. Continued 12

Compound	Cryst. cl. Space gr. Z	a, b, c	α, β, γ	Chromophore	Fe–L	Fe–Fe Fe–X–Fe	L–Fe–L	Ref.
$(NEt_4)_2[Fe_4(\mu_4\text{-}Sb)(CO)_{16}]$ (red)	m $P2_1/n$ 4	13.415(2) 19.601(2) 19.156(2)	90.973(9)	FeC_4Sb	OC 1.758(9,30) μ_4Sb 2.666(1,4)	not given Sb 109.5(-,1.0)	C,C 93.0(3,3.0) 119.8(4,2.5) C,Sb 87.0(2,1.8) 176.7(2,9)	393
$(PPh_4)_2[\{Fe(CO)_4\}(\mu_4\text{-}Te)]$ (dark purple)	m C2/c 4	14.264(4) 19.719(5) 22.937(10)	90.55(3)	FeC_4Te	OC 1.774(13.24) μ_4Te 2.640(2,7)	not given Te 109.5(1,8)	C,C 90.6(4,8) 120.0(5,1.5) C,Te 89.3(3,1.0) 179.0(4,2)	325
$[(OC)_{12}Fe_4(\mu_4\text{-}Se_2)_3](SbF_6)_2$. $3SO_2$ (orange red)	tr P-1 2	10.941(4) 18.400(9) 10.253(4)	93.10(4) 103.74(3) 93.98(3)	FeC_3Se_3	OC 1.813(9,3) μSe 2.415(2,16)	not given Se 87.78(6)	C,C 90.6(4,1.1) Se,Se 56.81(5,9)	395
$[Fe(\mu_3\text{-}AsMe)(CO)_3]_4.6C_6H_{12}$ (dark red)	c P-43n -	18.73(1)		FeC_3As_3	OC 1.78(3,3) μ_3As 2.412(4,6)	not given As 102.6(1,2)	C,C 96.3(12,5) As,As 75.9(1,1) C,As 93.1(9,8)	439
$Fe_4\{\mu_8\text{-}\eta^4\text{-}N_2C_8H_{10}N_2\}(CO)_{12}$ (yellow red)	m $P2_1/n$ 4	8.9036(6) 11.9657(7) 24.5507(11)	97.136(5)	FeC_3N_2	OC not given μN not given	2.502(1,2) not given	not given	231
$Fe_4\{\mu_8\text{-}\eta^4\text{-}Se_2C_4H_2Se_2\}(CO)_6$ (dark red)	m $P2_1/n$ 2	8.7764(2) 15.9229(6) 9.1431(10)	90.355(8)	FeC_3Se_2	OC 1.799(8,30) μSe 2.384(1,4)	2.5055(12) Se 63.45(3)	C,C 96.6(3,4.9) Se,Se 81.3(1,1) C,Se 96.3(2,8.6) 158.3(2,2.7)	440
$Fe_4\{\mu_8\text{-}\eta^4\text{-}Se_2C(Ph)C(H)Se_2\}$. $(CO)_{12}$ (purple)	tr P-1 2	8.913(1) 9.115(1) 17.556(1)	74.97(1) 84.96(1) 86.09(1)	FeC_3Se_2	OC not given μSe 2.368(2,23)	2.572(2) Se 64.5(-,2)	Se,Se 80.9(-,4)	191
$[Fe_4\{\mu_8\text{-}\eta^4\text{-}Se_2C(H)CCCMeSe_2\}$. $(CO)_{12}]C_4H_8O$ (dark red)	tr P-1 2	11.281(2) 11.952(2) 12.614(2)	88.69(2) 64.95(1) 67.10(2)	FeC_3Se_2	OC not given μSe 2.378(2,34)	2.537(2,8) Se 64.5(1,6)	Se,Se 81.1(1,4) Se,C 89.2(3) 163.6(4)	187

Table 7. Continued 13

$Fe_4\{\mu_8\text{-}\eta^4\text{-}Se_2C(Ph)C(H)Se_2\}$ $(\mu\text{-dppe})(CO)_{10}$ (violet)	tr P-1 2	12.969(1) 13.032(1) 18.669(1)	76.8(1) 72.74(1) 63.99(1)	FeC_3Se_2 (x2)	OC 1.793(15,38) µSe 2.402(2,13)	2.529(2,2) Se 63.8(1,7)	C,C 94.8(5,3.8) Se,Se 79.3(1,2) C,Se 98.4(4,12.6) 156.8(5,6.0)	441
				FeC_3Se_2P (x2)	OC 1.771(13,15) µSe 2.386(2,38) P 2.239(3,2)		C,C 93.7(5,2.1) Se,Se 79.8(1,1) C,Se 88.4(5,3.5) 153.9(5,7.6) C,P 92.8(4,1.7) Se,P 112.9(1,7.4)	

a Where more than one chemically equivalent distance or angle is present, the mean value is tabulated. The first number in parenthesis is e.s.d., the second is maximum deviation from the mean value.

b The chemical identity of coordinated atom/ligand is specified in these columns.

c There are two crystallographically independent molecules.

Table 8. Crystallographic and Structural Data for Pentanuclear Iron Carbonyls[a].

COMPOUND (COLOUR)	CRYST. CL. SPACE GR. Z	a[Å] b[Å] c[Å]	α[°] β[°] γ[°]	CHROMO-PHORE	Fe–L [Å]	Fe–Fe [Å] Fe-L-Fe [°]	L–Fe–L [°]	REF.
$(NMe_4)[Fe_5(\mu_5\text{-}N)(\mu\text{-}CO)_2(CO)_{12}]$ (black)	m $P2_1/m$ 2	9.973(2) 11.498(2) 12.401(2)	113.53(1)	FeC_3N (x2)	OC[b] 1.783(5,10) μ_5N 1.852(3) 1.931(3)	2.575(-,12) 2.6389(5) N[b] 87.4(1,4.2) 171.5(2,1.3) C 78.3(1)	C,C[b] not given	442
				FeC_3N (x2)	OC 1.777(3,3) μOC 1.852(3) μ_5N 1.834(1)			
				FeC_4N (x1)	OC 1.764(5,11) μOC 2.206(3,0) μ_5N 1.872(3)		C,C 179.67(9)	
$[N(PPh_3)_2][Fe_5(\mu_5\text{-}N)(\mu\text{-}CO)_2\cdot(CO)_{12}]$ (black)	m $P2_1/c$ 4	17.48(1) 15.99(2) 22.23(2)	123.93(4)	FeC_3N (x2)	OC 1.78(2,3) μ_5N 1.871(8,45)	2.569(3,19) 2.621(3,25) not given	not given	443
				FeC_3N (x3)	OC 1.77(2,6) μOC 1.90(2,4) 2.30(2,26) μ_5N 1.85(1,3)			
$Fe_5(\mu\text{-}H)(\mu_5\text{-}N)(\mu\text{-}CO)_3(CO)_{11}$ (brown)	m $P2_1/n$ 4	8.437(2) 15.056(3) 16.048(3)	95.42(2)	FeC_3HN (x2)	OC 1.92(1,4) μ_5N 1.836(3) 1.913(3)	2.58(2)	not given	412
				FeC_4N (x3)	OC 1.92(1,4) μ_5N 1.836(3) 1.9113(3)			

Table 8. Continued 2

(NO)].0.5(CH$_2$Cl$_2$ (not given)	P2$_1$/n 4	17.66(1) 17.97(1)	91.47(5)	(x4)	μOC 1.81(1,4) μ$_5$C 1.859(9,7) 1.975(9)	2.659(2,4) not given	90.5)-,2.4)	
				FeC$_4$N (x1)	OC 1,76(1) μOC 2.32(1,7) μ$_5$C 1.875(8) ON 1.73(1)		not given	
(NBu$_4$)$_2$[Fe$_5$)μ$_5$-C)(CO)$_{14}$] (black	m P2$_1$/c 4	12.719(3) 24.56(2) 21.68(2)	124.53(4)	FeC$_3$N (x4)	OC 1.77(2,9) μ$_5$N 1.86(2,2) 2.00(1)	2.553(3) 2.598(4,18) 2.675(4,5) not given	not given	443
				FeC$_2$N (x1)	OC 1.76(2,4) μ$_5$N 1.88(2)			
(ppn)$_2$[(OC)$_{13}$Fe$_5$(μ$_5$-C)(μ-SO$_2$)] (black)	m P2$_1$/c 4	19.384(5) 15.644(6) 26.612(4)	103.03(2)	FeC$_4$ (x3)	OC 1.76(1,1) μ$_5$C 1.89(1,6)	2.597(2) 2.657(2,17) C 88.5(4,3.1)	not given Fe,Fe 90.0(-,1.8)	305
				FeC$_3$S	OC 1.76(1,3) μ$_5$C 1.87(1,2) μS 2.167(3,3)	2.564(2) 2.638(2) C 87.7(4,1.3) S 72.6(1)	not given	
Fe$_5$(μ$_5$-C)(CO)$_{15}$ (not given)	m C2/c 8	15.135(8) 9.234(6) 32.97(3)	103.39(5)	FeC$_4$	OC 1.815(7,26) μ$_5$C 1.891(5,20) 1.953(5)	2.631(2,19) 2.687(2,27) not given	not given	443

Table 8. Continued 3

$(NBu^n_4)_2[Fe_5(\mu_3\text{-}S)_2(CO)_{14}]$[c] (black)	m $P2_1/c$ 8	33.284(7) 18.769(4) 18.704(4)	90.945(5)	FeC_2S_2 (x1)	OC 1.757(39,24) μ_3S 2.208(9,1)	2.632(6,6) S not given	S,S 90.9(3) S,Fe 52.6(2,2) 101.0(3,9) 143.3(3,1) Fe,Fe 58.4(2,1) 127.2(2,1.6) 154.8(2,9.3)	322
				FeC_3S (x4)	OC 1.72(4,16) μ_3S 2.172(10,13)	2.560(6,6)	S,Fe 53.9(3,2) Fe,Fe 60.8(2,1)	
				FeC_2S_2 (x1)	OC 1.794(44,34) μ_3S 2.196(10,11)	2.618(7,24) S not given	S,S 90.3(4) S,Fe 52.6(3,6) 102.5(3,2) 142.4(3,0) Fe,Fe 58.4(2,1) 126.3(2,2) 156.9(2,8.6)	
				FeC_3S (x4)	OC 1.75(4,19) μ_3S 2.166(10,22)	2.557(7,7)	not given	
$(ppn)_2[Fe_5(\mu_3\text{-}S)_2(CO)_{14}]$ (black)	m $P2_1/c$ 4	24.060(5) 14.355(6) 23.898(13)	90.42(3)	FeC_3S (x4)	OC 1.772(8,24) μ_3S 2.183(2,9)	2.560(1,4) 2.663(1,10) S 73.0(-,1.3)	Fe,Fe 60.0(-,1.2)	444
				FeC_2S_2 (x1)	OC 1.775(6,8) μ_3S 2.214(2,5)			

Table 8. Continued 4

$(PPh_4)_2[Fe_5(\mu_3\text{-}Se)_2(CO)_{14}]$ (dark brown)	tr P-1 2	12.951(4) 13.155(6) 20.792(8)	98.01(3) 94.25(3) 115.77(3)	FeC_2Se_2 (x1)	OC 1.777(20,7) μ_3Se 2.324(4,5)	2.649(5,9) Se 69.3(1,9)	Se,Se 86.8(1) C,Se 120.1(7,8.0) C,Fe 73.9(7,1) 94.4 126.9(7) Fe,Fe 58.6(1,1) 125.5(1,9) 159.2(1,5.2)	322
				FeC_3Se (x4)	OC 1.772(27,33) μ_3Se 2.307(3,33)	2.593(4,4)	C,Se 107.4(8) Se,Fe 55.5(1,8) Fe,Fe 60.7(1,3)	
$(NPr_4)_2[Fe_5(\mu_3\text{-}S)_4(CO)_{12}]$ (black)	m C2/c 8	34.16(1) 19.446(8) 17.902(8)	121.40(2)	FeC_3S_2 (x4)	OC not given μ_3S 2.311	2.483 S 65.2	S,S 104.9	445
				FeS_4 (x1)	μ_3S 2.342		S,S 82.0(2,1) 107.6(2,0) 146.9(2,2.2)	
$(NPr_4)[Fe_5(\mu_3\text{-}S)_4(CO)_{12}]$.0.5thf (black)	tr P-1 2	11.745(1) 11.867(1) 15.198(2)	68.58(1) 85.99(1) 75.46(1)	FeC_3S_2 (x4)	OC not given μ_3S 2.307	2.500 S 65.62	S,S 85.98	445
				FeS_4 (x1)	μ_3S 2.2682(8,9)		S,S 87.80(3,4) 121.3(1,2.3)	
$Fe_5\{\mu_8\text{-}\eta^4\text{-}N_2C_{11}H_{14}N_2\}(CO)_{15}$ (green)	m $P2_1/a$ 4	15.064(2) 14.749(2) 16.765(2)	116.29(1)	FeC_3N_2 (x5)	OC not given μN not given	2,521(1,16) 2.581(1,26)	not given	231
$Fe_5\{\mu_8\text{-}\eta^4\text{-}N_2C_{11}H_{16}N_2\}(CO)_{13}$ (black)	m $P2_1/c$ 4	9.223(5) 16.83(1) 19.58(2)	99.16(6)	FeC_3N (x4)	OC 1.778 μN 1.865(9,15)	2.539(3,16) 2.584(3,3) not given	Fe,Fe 60.0(1,1.2) 106.4 121.4(1)	446
				FeN_4C (x1)	μN 1.913(9,9) OC 1.778		N,N 42.3(4,3) 93.7(4,3.0)	

Table 8. Continued 5

$(NEt_4)_2[Fe_5(\mu_3\text{-}As)_2(CO)_{17}]$ (dark red)	tr P-1 2	12.466(2) 13.408(3) 15.457(3)	80.59(3) 72.01(3) 66.66(3)	FeC_3As_2 (x3)	OC 1.777(5,21) μ_3As 2.365(1,20)	2.714(1,15) As 69.9(-,6) 102.1(-,4)	C,C 99.0(2,9.6) As,As 77.1(1,5) C,As 91.8(1,5.7) 110.8(1,1.8) 156.8(1,8.8)	390
				FeC_4As (x2)	OC 1.777(5,21) μ_3As 2.400(1,12)		C,C 92.6(2,7.1) 115.9(2,5.2) 152.9(2,13.4) C,As 88.1(1,5.9) 172.8(2,6.6)	
$(NEt_4)_2[Fe_5(\mu_3\text{-}Sb)_2(CO)_{17}]$ (black)	tr P-1 2	13.443(3) 15.516(6) 12.610(4)	107.06(3) 112.84(2) 81.13(3)	FeC_3Sb_2 (x3)	OC 1.764(7,35) μ_3Sb 2.521(1,17)	2.842(1,18) Sb 68.0(-,5) 100.8(-,4) 130.7(-,4.3)	C,C 99.7(3,5.0) Sb,Sb 78.8(-,3) C,Sb 95.5(2,16.2) 156.6(2,7.9)	436
				FeC_4Sb (x2)	OC 1.769(7,17) μ_3Sb 2.529(1,11)		C,C 101.1(2,11.9) 129.0(2,5.9) 168.8(2) C,Sb 87.2(2,5.3)	

a Where more than one chemically equivalent distance or angle is present, the mean value is tabulated. The first number in parenthesis is e.s.d., the second is maximum deviation from the mean value.

b The chemical identity of coordinated atom/ligand is specified in these columns.

c There are two crystallographically independent molecules.

Table 9. Crystallographic and Structural Data for Hexanuclear Iron Carbonyls[a].

COMPOUND (COLOUR)	CRYST. CL. SPACE GR. Z	a[Å] b[Å] c[Å]	α[°] β[°] γ[°]	CHROMO- PHORE	Fe–L [Å]	Fe–Fe [Å] Fe-L-Fe [°]	L–Fe–L [°]	REF.
$(PPh_4)_2[Fe_6(\mu_6\text{-}N)(\mu\text{-}CO)_3\cdot(CO)_{12}(H)]$ (black)	m $P2_1/n$ -	19.757(4) 14.866(3) 23.056(4)	 113.64(2)	FeC_3N (x6)	OC[b] 1.748(11,63) μOC 2.004(10,94) μ_6N 1.862(5,15) H not located	2.579(2,31) 2.661(2,24) not given	C,C[b] not given	447
$(NEt_4)_2\{NMe_2(CH_2Ph)_2\}_2[Fe_6(\mu_6\text{-}N)(\mu\text{-}CO)_3(CO)_{12}]$ (black)	m $P2_1/c$ -	12.299(6) 21.576(7) 21.153(5)	 102.70(3)	FeC_3N (x6)	OC 1.749(8,19) μOC 1.941(7,39) μ_6N 1.864(4,13)	2.545(1,12) 2.665(1,59)	not given	447
$(PPh_4)_2[Fe_6(\mu_6\text{-}C)(\mu\text{-}CO)_4\cdot(CO)_9(NO)_2]$ (black)	m C2/c 4	37.05(1) 12.635(5) 18.728(6)	 137.04(3)	FeC_5 (x2)	OC 1.77(1,1) μOC 1.94(1) 2.56(1) μ_6C 1.884(1)	2.594(3,5) 2.680(2,40)	Fe,Fe 60.0(-,3.2) 90.0(-,1.8)	307b
				FeC_3N (x2)	OC 1.96(1) μOC 1.98(1) μ_6C 1.90(1) ON 1.65(1)		not given	
				FeC_4 (x2)	OC 1.76(1,2) μOC 1.82(1) μ_6C 1.86(1)		not given	
$\{N(Me_3)PH_2Ph\}_2[(OC)_{14}Fe_6(\mu\text{-}CO)(\mu_6\text{-}C)(\mu\text{-}SO_2)]$ (black)	or $Pna2_1$ 4	14.642(2) 15.142(2) 19.329(3)		FeC_4 (x4)	OC not given μ_6C 1.87(1,2)	2.626(2,44) 2.664(2,6) C 88.6(5,1.6)	not given Fe,Fe 90.0(-,1.4)	305
				FeC_3S (x2)	OC not given μ_6C 1.89(1,0) μS 2.180(3,6)	2.675(2,39) C 89.7(5,1.3) 74.45(9)	not given	

Table 9. Continued 2

$(NEt_4)_2[Fe_6(\mu_5\text{-}Sb)_2(CO)_{20}]$ (brown)	or $P2_12_12_1$ 4	15.962(6) 25.878(6) 10.035(2)		FeC_3Sb_2 (x4)	OC 1.761(1,21) μ_5Sb 2.521(2,7) 2.672(2,30)	2.832(2,31) Sb 66.0(-,1.2) 100.6(-,4.1) 129.7(-,4.0)	C,C 96.4(6,3.4) Sb,Sb 79.4(-,7) C,Sb 90.3(4,8.5) 133.7(4,1.2) 161.3(4,1.6)	437
				FeC_4Sb (x2)	OC 1.74(2,4) μ_5Sb 2.538(2,0)		C,C 91.9(6,3.4) 119.9(7,6.0) C,Sb 88.2(6,4.7) 174.4(5,6)	
$[Fe_6(\mu_3\text{-}PMe)_2(\mu_4\text{-}P)_2(CO)_{18}]$ (deep purple)	m Cc 8	16.299(6) 8.779(4) 22.619(7)	105.55(3)	FeC_3P_2 (x6)	OC not given μ_3P 2.278(5,55) μ_4P 2.238(5,50)	2.714(4,22) 3.591(4,9)	not given	448
$[Fe_3(\mu\text{-}CO)_2(CO)_9\{\mu\text{-}dpph\}]_2$ (dark green)	tr P-1 1	12.190(15) 14.908(8) 9.209(6)	95.56(5) 112.29(7) 108.37(8)	FeC_4P (x2)	OC 1.795(8,8) μOC 1.856(7) 2.153(8) P 2.278(2)	2.593(1) 2.691(2,11) C 77.2(2,2.2)	C,C 89.6(3,10.0) 171.4(3,2.2) C,P 93.1(2,6.3) Fe,Fe 60.0(-,2.4)	449
				FeC_5 (x2)	OC 1.807(10,21) μOC 1.895(7) 2.348(6)		C,C 91.4(4,9.3) 172.6(4,3.2)	
				FeC_4 (x2)	OC 1.791(9,8) 1.829(8)		C,C 91.8(4,9) 101.0(4)	
$(NPr_4)_2[Fe_6(\mu_3\text{-}S)_4(\mu\text{-}S)_2(CO)_{12}]$ (black)	or $Cmc2_1$ 4	18.870(8) 17.307(13) 16.430(3)		FeC_3S_2 (x4)	OC not given μ_3S 2.307	2.504 S 65.80	S,S 83.6	445
				FeS_4 (x2)	μ_3S 2.304(4) μS 2.210(4)		S,S 104.9	

Table 9. Continued 3

(NEt$_3$Bz)$_2$[Fe$_6$(μ$_3$-S)$_4$(μ-S)$_2$.	tr	8.025(7)	68.43(4)	FeC$_3$S$_2$	OC not given	2.499(1)	S,S 86.15(6,61)	450
(CO)$_{12}$]	P-1	12.956(8)	72.15(5)	(x4)	μ$_3$S 2.321(2,9)	μ$_3$S not given		
(purple)	1	13.676(7)	89.44(4)					
				FeS$_4$	μS 2.200(1,3)	2.693(1)	S,S 104.52(1)	
				(x2)	μ$_3$S 2.305(2,18)	μS 75.48(5)		
(NEt$_4$)$_2$[Fe$_6$(μ$_3$-S)$_4$(μ-S)$_2$(CO)$_{12}$]	m	14.037(4)		FeC$_3$S$_2$	OC 1.779(11,25)	2.504(2)	S,S 84.7(1)	322
(red)	P2$_1$/n	12.053(4)	112.51(2)	(x4)	μ$_3$S 2.306(2,3)	μ$_3$S 65.8(1,1)	C,S 107.6(2)	
	2	14.834(4)						
				FeS$_4$	μS 2.206(2,2)	2.681(2)	S,S 85.5(1)	
				(x2)	μ$_3$S 2.290(2,2)	μS.84.8(1,1)	105.2(1)	
							116.5(1,5)	
(ppn)$_2$[Fe$_6$(μ$_3$-S)$_4$(μ-S)$_2$(CO)$_{12}$]	m	34.424(4)		FeC$_3$S$_2$	OC 1.778(10,20)	2.506(1)	C,C 97.4(4,3.5)	444
(black)	C2/a	14.081(2)	115.72(1)	(x4)	μ$_3$S 2.301(2,7)	μ$_3$S 79.2(1,13.3)	S,S 84.1(1,1.1)	
	4	19.674(2)					C,S 95.1(3,11.4)	
							156.8(3,2.0)	
				FeS$_4$	μ$_3$S 2.305(2,5)	2.675(2)	S,S 104.96(7)	
				(x2)	μS 2.198(2,1)	μS 74.98(7)	116.9(1,2.6)	
(NMe$_3$PH$_2$Ph)$_2$[Fe$_6$(μ$_3$-Se)$_4$.	tr	9.687(2)	103.47(2)	FeC$_3$Se$_2$	OC not given	2.552(1)	not given	451
(μ-Se)$_2$(CO)$_{12}$]	P-1	10.540(3)	95.57(2)	(x4)	μ$_3$Se 2.419(1,8)	μ$_3$Se 63.69(4,15)		
(dark purple)	1	12.936(3)	111.00(2)					
				FeSe$_4$	μSe 2.320(1,2)	2.750(2)	S,S 85.84(4)	
				(x2)	μ$_3$Se 2.412(1,4)	μSe 72.72(4)	107.28(5)	
							116.41(5,1.84)	

Table 9. Continued 4

[$Fe_6(\mu_4\text{-}S)_2(\mu\text{-}CNEt_2)_2(CO)_{18}$].). $5CH_2Cl_2$ (dark red)	m C2/c 8	46.754(2) 9.268(1) 19.628(2)	100.7(1)	FeC_4S (x4)	OC 1.80(2,11) μNC 1.90(2,2) μ_4S 2.268(4,12)	2.479(3,0) C 81.4(6,2) S 66.2(1,1)	C,C 96.5(7,6.3) C,S 91.8(5,3) 106.6(5,1.1) 156.5(5,1.4)	452
				FeC_3S_2 (x2)	OC 1,77(2,3) μ_4S 2.255(4,4)	2.544(3) S 68.7(1)	C,C 95.9(8,9.4) S,S 82.5(1,1) C,S 96.8(5,10.2) 156.6(6,3.2)	
{[$Fe_2(CO)_8(\mu_4\text{-}As)]_2[Fe_2(CO)_6$]} (green black)	or Pbca 8	22.908(7) 17.521(9) 16.071(7)		FeC_3As (x2)	OC not given μ_4As 2.34(1,2)	2.72(1,4) μ_4As 69.7(3,5)	not given	453
				FeC_4As (x4)	OC not given μ_4As 2.41(1,2)		not given	
[$Fe_6(\mu\text{-}SC_6H_9)_2(\mu_4\text{-}S)_2(CO)_{18}$] (dark red)	m $P2_1/c$ 4	14.319(9) 10.901(9) 27.139(20)	92.59(5)	FeC_3S_2 (x4)	OC 1.77(2,3) μ_4S 2.273(5,3) μS 2.267(5,12)	2.528(3,33) μ_4S 68.1(1,1.2) 133.2(1,9.5) μS 67.3(1,5)	S,S 76.0(2,1) C,C 95.7(8,5.9) S,C 91.8(9,1.4) 105.3(9,5.1) 155.8(6,1.9)	454
				FeC_3S_2 (x2)	OC 1.75(2,2) μ_4S 2.259(4,13)		S,S 82.4(2,0) C,C 95.7(8,6.5) S,C 89.5(6,9) 104.4(6,2.6) 156.9(6,6.2)	

a Where more than one chemically equivalent distance or angle is present, the mean value is tabulated. The first number in parenthesis is e.s.d., the second is maximum deviation from the mean value.

b The chemical identity of coordinated atom/ligand is specified in these columns.

Table 10. Crystallographic and Structural Data for Hepta- and Octanuclear Iron Carbonyls[a].

COMPOUND (COLOUR)	CRYST. CL. SPACE GR. Z	a[Å] b[Å] c[Å]	α[°] β[°] γ[°]	CHROMO-PHORE	Fe–L [Å]	Fe–Fe [Å] Fe-L-Fe [°]	L–Fe–L [°]	REF.
$(AsPh_4)_2[Fe_7(\mu_7\text{-}B)(\mu\text{-}CO)$. $(CO)_{19}(H)]$ (black)	m $P2_1/c$ 4	11.248(2) 14.336(2) 41.698(6)	 90.30(1)	FeC_3B (x2)	OC^b 1.763(9,11) μOC 1.857(8,13) μ_7B 2.027(8,6)	2.563(2,4) 2.636(2,10) 2.715(2,43) C^b 81.5(3,4) B 78.6(3,7.7) 138.0(4,8.4)	C,C^b 92.1(4,3.8) C,B 115.0(4,2.7) 140.3(4,1.3)	455
				FeC_3B (x4)	OC 1.779(9,28) μ_7B 2.091(9,56)		C,C 93.7(4,5.4) C,B 106.1(4,1.4) 129.7(4,3.5)	
				FeC_4B (x1)	OC 1.760(8,4) μOC 2.066(8,4) μ_7B 2.356(8)		C,C 88.3(4,2.6) 172.4(3) C,B 93.8(3,8) 134.5(5,1.0)	
$[\{Fe_4(\mu_3\text{-}Te)_4(CO)_{10}\}_2$. $(\mu_4\text{-}\eta^2\text{-}Te_2)]$ (black)	tr P-1 2	14.930(4) 24.070(3) 13.045(2)	93.19(1) 113.36(1) 76.61(2)	FeC_3Te_3 (x4)	OC 1.78(-,7) μ_3Te 2.631(2,26)	2.619(1,4) Te 96.5(-,5.5) 121.7(-,1.2)	Te,Te 82.1(-,2.4) C,C 97(-,6)	428
				FeC_2Te_3 (x4)	OC 1.78(-,7) μ_3Te 2.607(2,23) μ_4Te_2 2.620(2,21)		Te,Te 84.5(-,3.6) 97(-,6)	
$(NMe_3CH_2Ph)_2[Fe_8(\mu_3\text{-}Te)_6$. $(CO)_{24}]$ (dark red)	m C2/c 4	20.694(7) 14.649(4) 22.813(7)	 101.45(3)	FeC_3Te_2 (x6)	OC not given μ_3Te 2.566(3,18)	2.601(3,24) not given	Te,Te 77.0(1,2.0)	224 456
				FeC_3Te_3 (x2)	OC not given μ_3Te 2.636(3,3)		Te,Te 72.67(7) 91.08(8) 162.92(11)	

Table 10. Continued 2

[{$Fe_4(\mu_4\text{-}PPh)_2(CO)_{10}\}_2$· {$\mu$-$(MeO)_2PC_6H_4P(OMe)_2$}] (not given)	tr P-1 1	11.355(6) 10.058(6) 16.152(7)	103.78(7) 91.38(6) 68.34(6)	FeC_3P_2 (x6)	OC not given μ_4P 2.30(2,6)	2.535(2,18) 2.727(1,30) not given	not given	457
				FeC_2P_2 (x2)	OC not given μ_4P 2.30(2,6) P 2.169(2)			

a Where more than one chemically equivalent distance or angle is present, the mean value is tabulated. The first number in parenthesis is e.s.d., the second is
maximum deviation from the mean value.

b The chemical identity of coordinated atom/ligand is specified in these columns.